Contents

Simply Strawberries

by Sara Pitzer

GARDEN WAY PUBLISHING
POWNAL, VERMONT 05261

Illustrations by Elayne Sears
Book Design by Andrea Gray
Cover Photo by Didier Delmas

Thanks to the Carolina Department of Agriculture, Chadbourn Extension Homemakers'
Club, California Strawberry Advisory Board, Florida Department of Agriculture, Florida
Strawberry Growers' Association, and the U.S. Department of Agriculture for ideas and
information.

© **Copyright 1985 by Storey Communications, Inc.**
The name Garden Way Publishing has been licensed to Storey Communications, Inc. by Garden Way, Inc.

Printed in the United States by Capital City Press
First printing, March, 1985

Library of Congress Cataloging in Publishing Data

Pitzer, Sara.
  Simply strawberries.

  1. Cookery (Strawberries) I. Title.
TX813.S9P58   1985   641.6'475   84-28790
ISBN 0-88266-383-6
ISBN 0-88266-382-8 (pbk.)

Introduction

uring one of the Sunday services when I was a child, a visiting minister remarked that he was thoroughly sick of hearing about Heaven's pearly gates and golden streets. Heaven, he said, was a personal award that he visualized as a land where, among other pleasant happenings, juicy red strawberries ripened eternally. It was one of the best sermons I'd ever heard.

Lewis Hill
FRUITS AND BERRIES
FOR THE HOME GARDEN

In my entire life I have known only one person who didn't like strawberries. My father claims that having earned his pocket money as a boy by picking strawberries for two cents a quart has spoiled his taste for them forever. Poor Daddy. Forever seems like such a long time to do without strawberries. And he grew up with the best. My Grandmother Dietrick created a wonderful hybrid strawberry that she called "Pocono-Pocahontas" or "Sadie's Pocahontas"—or some such name, coined to let you know that part of the credit went to an old reliable variety and the rest of the credit went to her. It wasn't the biggest strawberry around, but its flavor was so intense it nearly made your throat ache. You could put a whole berry in your mouth comfortably and bite down on it to release a rush of incredibly sweet juice.

That's always been my favorite way to eat a strawberry. Still is. But over the years I've discovered a plethora of other ways to serve strawberries to satisfy the craving for variety we all seem to bring to meal time. The very first variation I remember was another product of my grandmother's imagination. She called it "Teapot Juice" because she mixed it up for early summer tea parties my sister, Carolyn, and I always had with Raggedy Ann and Andy and Baby Doll Katherine, and sometimes Cousin Katherine, too. Teapot Juice was basically lemonade with a goodly portion of mashed strawberries and their juice added. She didn't bother to puree the berries, just mashed them with a fork. The bits of strawberry came out of the spout of the teapot and settled in the bottom of the teacup so that you had to finish with a spoon.

I have included a recipe for Teapot Juice in this collection, imitating Grandma's as nearly as I can. To me it doesn't taste quite as good as hers, but perhaps that's because I've been tasting it without the magic of Raggedy Ann and Andy and Carolyn and Baby Doll and Cousin Katherine at the table.

The tea party magic may escape me, but the magic of strawberries endures, for me and for everybody. (Except poor Dad.) Watch what happens in a restaurant when people see strawberries on a menu. Somebody always says, "Oh, look—strawberries," in that light and delighted voice that tells you they mean to have some.

Given that knowing something is good for them spoils food for some people, I hesitate to mention this, but strawberries are an uncommonly rich source of vitamin C, providing more of the vitamin in a cupful than you would get from a medium orange. (Heating, freezing, or storing bruised berries destroys much of the vitamin C.) Strawberries are also a good source of iron. They contain only about sixty calories a cup. But that's the kind of information you can enjoy knowing at the same time everyone else is simply enjoying good taste. It's a good feeling when you serve strawberries at home, and everybody at the table interrupts conversation for a moment of pleased "ah-h-h-s."

In the grocery store, the early, high-priced boxes of strawberries from California and Florida are nearly irresistible, even though you know that the local berries that come a little later will be better.

And when you finally get your hands on strawberries fresh from the patch, still a little warm from the sun, we're back to the magic of putting Sadie's whole Pocono-Pocahontas berries into your mouth all over again.

My father grew berries for the rest of us, in spite of being indifferent to them himself. Having grown up with an abundance of strawberries from the garden in a family full of highly individualistic cooks, I learned that there are enough ways to create such classic favorites as strawberry preserves and strawberry shortcake to fill an entire book just with variations on a few basic recipes. The amazing thing is how hard it is to declare a personal preference, let alone judge one version actually better than another. For instance, Grandmother

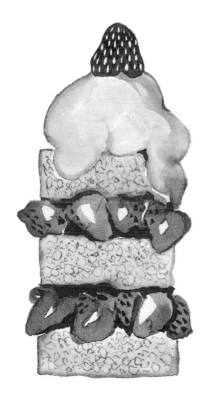

Dietrick made strawberry preserves without commercial pectin. The berries floated in the syrup, which was always thin enough to run off the edges of the bread down between your fingers. Good. Grandmother Pennington, on the other hand, added commercial pectin to her preserves and mashed the berries slightly so that the preserves spread evenly across the toast. Good. The best a kid could hope for was to spend equal time in both households and never have to choose one over the other. I still can't choose. I make both. The recipes are in this book.

Strawberry shortcake gives me even more trouble. I remember eating a shortcake made of warm biscuit seasoned with nutmeg, spread with butter, and covered with cold strawberries which had been mashed with sugar to release the juice. We poured heavy cream over everything and added a few whole berries. Usually that was all we ate for supper. Nothing could have pleased us more than the contrasts—hot, crisp, buttered biscuits, cold, sweet-tart berries with sugar, and rich cream, thick and liquid. But I remember also a concoction of strawberries and homemade sponge cake engulfed in mountains of real whipped cream touched with vanilla. Light and heavenly. We called that strawberry shortcake, too. Who can choose? Not me. I've included recipes for both, along with an assortment of others that people I know consider their favorites.

This kind of talk makes it sound as though the world revolves around desserts and preserves alone. It would be a shame to limit the presence of strawberries to morning toast and the end of meals. I am especially proud of the assortment of recipes in this book for salads and light entrées because I think it's time to explore all the ways the flavor of strawberries complements other foods. All kinds of new possibilities are coming to light. For example, in my experimenting I discovered that strawberries go beautifully with seafood—especially shrimp and salmon. I've noticed that professional caterers are combining strawberries with salad greens and garnishing deviled eggs with strawberries. The strawberry contests sponsored by growers and extension agents across the country are producing novel ways to bake strawberries in bread and turn them into spreads.

And now that home freezers and a thriving commercial strawberry packaging industry make it possible for us to have high quality frozen strawberries any time of the year, I have tried to find as many ways as possible to use frozen strawberries. Recipes suitable for using both sweetened and unsweetened frozen strawberries are so marked throughout. Also, I hope the directions for freezing strawberries (page 9) will encourage you to put up your own supply at home during the relatively brief period in early summer when the supply is plentiful.

SELECTING AND CARING FOR STRAWBERRIES

If you've grown your own strawberries, selecting them is mainly a matter of picking them as they ripen. If you've ever seen a little child picking berries, you know that just grabbing the berry and pulling it from the plant is the worst thing to do in a berry patch. The hull stays on the stem, the berry gets gooey where the hull pulled out, and dirt sticks to the goo. Or, the hull stays on the berry and the pulling breaks other berries off the plant or heaves some of the roots out of the ground. To pick strawberries, use both hands. In one hand hold the stem close to the plant; in the other, hold the stem as close as you can to the berry. Twist and pull, leaving as short a stem as possible attached to the berry.

Pick strawberries when they are fully ripe, unless you have to keep them more than a day or so, in which case you should pick berries still a little pink. But remember that no strawberry picked before it is fully ripe will ever have the flavor of a plant-ripened one. Although a strawberry may deepen in color after it's picked, it doesn't really ripen—no more sugar is formed in the berry. If you have your own strawberries it's better to pick every couple of days than to pick unripe berries; if you visit pick-your-own farms, select only ripe berries and return a second time rather than trying to save some berries from a single picking.

It is better to pick strawberries in the morning before the sun has had time to soften them. Although it seems easier to carry a large container and fill it, strawberries keep better if you put them into small baskets so that the bottom layers aren't crushed under the weight of more berries on top.

Strawberries should be cooled (but not washed) as soon as possible after they are picked. A refrigerator is not necessarily the best place to cool them. The Amish in central Pennsylvania, where I used to live, cooled their berries in spring houses, special sheds built under shade trees, and even in the darkened pantries of their houses. I kept my own strawberries on the inside back steps of our old farmhouse, where they were insulated from the outside heat by a couple feet of masonry and plaster and from the inside by a wall of kitchen cupboards. If you have a comparable spot, a cellar or well-insulated closet perhaps, try it out. Strawberries hold up a bit better at cool temperatures not quite as cold as those in the refrigerator.

If you are planning to visit a pick-your-own farm for strawberries, find out their schedule. Some places have picking only every other day to give the berries a chance to ripen. Whatever the schedule, try to be there ahead of the hordes. I have been to a number of different pick-your-own places at various times of the day and on various days of the week in Pennsylvania, California, North Carolina, and South Carolina. I've concluded that Friday afternoon and all day Saturday are the most popular picking times and that early Monday morning is probably the *best* picking time, since the farms usually aren't open Sundays (which gives the berries an extra day to ripen) and since hardly anybody but me ever gets there early on a Monday.

If you must buy your strawberries, I think the best quality usually comes from local farmers and produce stands, commercial farmers' markets, small, high-quality grocery stores, and supermarkets—in that order. You can't do a lot of choosing when you buy berries by the basket. It's like trying to choose the best bushel of apples or the best sack of potatoes. The good and the bad are all mixed in there together. To the degree that you can choose, select strawberries that are firm, plump, and bright red, with a natural shine. The caps should still be intact and look bright green and fresh. Strawberries that are more orange than red, have little soft spots on them, or seem softish and wilted aren't going to taste any more appetizing than they look.

When you get your strawberries home, take them out of the basket or box in which you carried (or purchased) them and spread them in a single layer in shallow pans until you are ready to use them. Keeping them in a single layer is helpful enough in preserving strawberries that some grocery stores are now packaging them that way in styrofoam trays.

Wash strawberries just before you use them. Leave the hulls on until *after* you have washed the berries. The closer to serving time strawberries are washed and hulled the more flavor and nutrients you will retain.

WEIGHTS AND MEASURES

Being specific in measurements for strawberries is difficult because the size of the berries affects how many will fit into a cup. A cupful of big strawberries contains a little less actual strawberry than a cupful of small strawberries because of the way they fit into the cup. Weights are a much more accurate way to keep track, but most of us are accustomed to recipes in which ingredients are measured by cups. Moreover, scales are not common kitchen equipment for many of us. Therefore, I have used cups as the base measure for strawberries in my recipes. I suggest that you try to have a few extra berries on hand in case a pie or salad comes up looking a little lean. Similarly, it's fine to use fewer berries than the recipe calls for if there seem to be too many.

FREEZING AND PRESERVING

It's not an unnatural impulse to want to have something as spring-like as strawberries in the bleakness of winter. Housewives used to can them; make jams, jellies, and preserves from them; and even dry them as a fruit "leather". I offer recipes for jams, jellies, and preserves here, but I haven't talked about canning strawberries because I don't think it's worth the trouble or the sacrifice of perfectly good strawberries. I thought I was going to have to defend that point on the strength of my expertise alone, until I saw some canned strawberries in the grocery store, put out by a company that specializes in canning fruits and berries. On the back of the lable I read, "We hate to apologize for strawberries—but . . . " The copy went on to explain that even the firmest berries (which don't have the best taste) get mushy, turn brownish, and shrink in canning. What's left is lots of strawberry juice. The label copy concludes that there are lots of nice uses for strawberry juice. There are. But not enough to warrant canning strawberries now that it's possible to freeze them. In freezing, the flavor and color remain pretty much as they were fresh. You do lose texture. A frozen berry, completely thawed, is mushy. That's why commercial food packagers recommend that you serve their frozen strawberries before they've completely thawed. I advise the same for home-frozen strawberries.

You can freeze strawberries whole or sliced, mashed or pureed, unsweetened or in sugar or syrup. Here's how.

Whole, unsweetened. I consider this the most useful way to freeze strawberries. Wash, hull, and drain the berries. Place them in a single layer on cookie sheets in the freezer. The temperature should be as cold as you can get it to freeze the strawberries rapidly. When the strawberries are hard, quickly put them into freezer-quality plastic bags, forcing out as much air as possible, and store in the freezer.

Whole, sweetened. Allow about 3/4 cup sugar for each 5 cups of strawberries. Wash, hull, and drain the strawberries. Gently mix the sugar with the berries, spread them in a single layer on cookie sheets in the freezer. Freeze, package, and store as for unsweetened whole strawberries.

Sliced, sugar pack. Allow 1 cup sugar for each 6 cups sliced strawberries. Mix the sliced strawberries with the sugar and allow to stand until the sugar dissolves, about 10 minutes. Package in plastic freezer containers, leaving ½ inch headspace. Yield: About 3 pints for each 6 cups of berries.

Sliced or whole, syrup pack. For each 6 cups whole strawberries dissolve 1 cup sugar in 2 cups water. Bring to a boil. Chill. Slice the berries or leave them whole. Pack them lightly into plastic freezer containers and cover with the cold syrup, leaving ½ inch headspace. Yield: about 3 pints for each 6 cups strawberries.

Whole, water pack, unsweetened. This method yields only fair results. I don't recommend it. About the only use I can see for berries frozen this way would be to put the frozen blocks in a punch bowl. Pack strawberries into plastic freezer containers and cover with water containing 1 teaspoon crystalline ascorbic acid to each 4 cups water. Leave ½ inch headspace.

Mashed or pureed, sweetened or unsweetened. Wash, hull, and mash or puree the strawberries. Add sugar to taste, if you wish. Pack into *small* plastic freezer containers, or pour about ½ cupful into each pint freezer bag, or pour into ice cube trays. If you use either of the first two options, seal and freeze at once. If you use ice cube trays, freeze and then remove the cubes from the trays and store them in plastic freezer bags. Mashed and pureed berries should be frozen in small packages because larger ones take too long to freeze and to thaw.

USING FROZEN STRAWBERRIES

Although I have indicated the recipes in this book which I find suitable for using frozen strawberries, you may want to try them in others as well. As a general rule, I find frozen strawberries are adequate in any recipe using pureed berries or in any recipe that directs you to cook the berries, providing you can reconcile the sugar requirements of the recipe with the amount of sugar in your frozen package. That's why I prefer whole frozen unsweetened berries. It's easy enough to slice, puree, or mash them and to add sugar, but it's impossible to reverse any of those acts for a recipe using whole, sugarless berries. And while I recommend fresh strawberries for most shortcake recipes, it would be entirely possible to improvise good fillings for the cakes from sliced and sweetened berries or from unsweetened berries frozen whole to which you add sugar shortly before serving.

In addition to using frozen strawberries in recipes, don't overlook such uses as dropping whole frozen strawberries into punch bowls and tall drinks instead of ice cubes to chill without diluting the beverage. Puree frozen into cubes can be used the same way. Frozen berries buzzed in the blender with milk, juice, or yogurt create a "frozen" drink lower in calories than milkshakes. Also, don't overlook the uses for a puree of frozen berries (this is different from berries made fresh and then frozen) as an appetizer ice, garnish for fruit salads, or sweetened a bit to make a low-calorie dessert.

If you buy commercially frozen strawberries, you will find that whole, unsweetened strawberries almost always come in 16-ounce packages. These contain about 3 cups of berries, measured whole and frozen. That makes about 2 cups of puree. All the packages of sliced and sweetened frozen strawberries I have seen contained 10 ounces. This makes just over a cup, depending on how much of the syrup you keep. (Some brands contain more liquid than berries.)

A LITTLE GILDING
FOR PERFECT STRAWBERRIES

When your strawberries truly are perfect, picked from a well-maintained patch at peak ripeness, the simplest serving ideas are the best. A ripe strawberry grown in good soil, of a variety bred for flavor rather than size or shipping stamina, is sweet enough to need little or no sugar and juicy enough to make sauce superfluous. Supermarket strawberries, off-season imports, and late-season stragglers are better used in the more elaborate recipes in which additional ingredients combine with the strawberries to make more complex flavor concoctions. The following suggestions are for perfect berries only.

- Leave the hulls on the strawberries and serve them with small dishes of brown sugar and sour cream for dipping.

- Leave the hulls on the strawberries, dip each one into lightly beaten egg white, roll in granulated sugar, and set aside to dry for a few minutes before serving.

- Hull the strawberries and roll them in a mixture of finely chopped nuts and sugar.

- Fold whipped cream into sour cream, put a small mound in the center of individual serving plates, and arrange perfect whole berries with the hulls on, around the cream.

- Fill individual dessert glasses with whole, hulled strawberries. Puree an 8-ounce package of frozen red raspberries. Strain the puree and spoon a small amount over the strawberries before serving.

- Marinate whole, hulled strawberries in Grand Marnier for 30 minutes before serving.

- Fill individual dessert glasses with whole, hulled berries. Put a small spoonful of strawberry sherbet or ice on top of each serving and pour a generous splash of cold champagne over all.

- Melt 2 cups of chocolate morsels and ¼ cup unsalted butter over hot water. Dip each whole, unhulled strawberry into the chocolate, then chill 15 minutes before serving.

- Fill individual dessert glasses with whole, hulled berries. Simmer 1 cup port wine with 1 cinnamon stick, 2 lemon slices, and 1 whole clove. Chill. Pour over the strawberries just before serving.

- Puree a handful of less-than-perfect strawberries. Flavor with a spoonful of strawberry liqueur. Use as a sauce to spoon over the perfect whole berries in dessert dishes.

Stuffed Strawberries

YIELD: 6 SERVINGS AS AN
    APPETIZER

1 3-ounce package cream cheese
⅓ cup finely chopped nuts
2 tablespoons honey
½ pint whole strawberries
Watercress for garnish (optional)

Inspired by the old standard, stuffed prunes, these are a whole lot prettier. They're nice for buffets, lunch boxes, and garnishes. Look for large berries. They'll hold more stuffing.

Wash the berries, leaving caps on, and drain them on a towel. Cut a small wedge out of the side of each berry. With a fork or in the food processor mash together the wedges cut from the berries, the cream cheese, chopped nuts, and honey. Stuff the mixture into the cuts in each berry. Cover a serving tray with watercress and arrange the berries on it. Substitute another green if you can't find watercress, but it is preferable because of its deep color, contrasting bite, and small leaves, which help hold the berries in position.

STRAWBERRY PUREE

Several recipes that follow call for strawberry puree. In addition, it is good by itself and as a base for your own inventions. Making it is simple. Use the ripest berries you can find and either force them through a food mill or sieve or puree them in a blender or food processor. Sweeten to taste with honey or granulated sugar.
1 pint strawberries = 2 ½ cups sliced berries = 1 ⅔ cups puree.

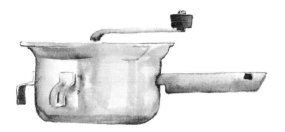

FROZEN STRAWBERRY PUREE

Unsweetened frozen strawberries, thawed for about 15 minutes and pureed in a food processor or blender, make a delightful, easy dessert. If you process just until the puree is coarse and icy, then sweeten to taste, the result is like strawberry ice. Process longer, until the puree is smooth and creamy, and the result tastes more like sherbet.

This puree is good served on pancakes and French toast or with blintzes—the iciness of the puree contrasting nicely with the hot pancakes, toast, or blintzes.

If you use commercially frozen whole strawberries, a 1-pound package makes about 2 cups frozen puree.

Strawberry Leather

Strawberry leather is a camper's and hiker's classic because it is light to carry and doesn't need refrigeration. Kids who aren't hiking any farther than the back porch like it, too, because it tastes so good.

Caution: be careful when eating leather. It's so good that you're apt to forget how concentrated dried fruits are and eat too much.

5 cups strawberries
¼ cup sugar

Puree the strawberries a cup at a time in the blender or food processor. Add the sugar and process until it dissolves.

Line two 15½ by 10½ inch jelly roll pans or cookie sheets with plastic wrap. Pour the puree evenly into the pans and place in a home dehydrator or a 150° F. oven to dry. (Be sure the oven never gets any hotter or it will brown the leather and spoil it.) Turn the pans every 2 hours. The leather should be dry and not sticky in about 5 hours. Remove it from the oven, peel off the plastic wrap, and cool. To store, wrap in fresh plastic and roll up like a jelly roll.

The leather will keep for about a month at room temperature and for up to 3 months in the refrigerator. The Florida Strawberry Growers Association says it will keep for a year in the freezer but I've never had any around that long. People eat it too fast.

Soups

North American Indians used to go on "veritable strawberry sprees, eating the delicate berries, seasoning their meat with them, drinking strawberry soup or a tea made from the leaves."

Virginia Scully
*TREASURY OF AMERICAN
INDIAN HERBS*

Strawberry Soup

SUITABLE FOR UNSWEETENED
  FROZEN STRAWBERRIES
YIELD: 6 SERVINGS

2 cups sliced strawberries
1 cup milk
2 tablespoons instant nonfat dry
  milk
1 cup light cream
2 tablespoons honey (more or less
  to taste)
1 teaspoon cinnamon
Whole strawberries for garnish

The hotel in Florida where I first tasted strawberry soup served it as an appetizer. I think it works equally well for dessert. The chef "respectfully declined" when I asked for his recipe. Here's my own version.

Puree the strawberries in a blender or food processor. Gradually add the milk, dry milk, cream, honey, and cinnamon, processing continuously. Chill before serving unless all the ingredients were very cold when you made the soup. Garnish each serving with several whole berries.

Strawberry-Port Soup

SUITABLE FOR UNSWEETENED
 FROZEN STRAWBERRIES
YIELD: 6 SERVINGS

4 cups sliced strawberries
3 cups port wine
1 cup water
¼ cup sugar
2 tablespoons arrowroot
¼ cup water
6 tablespoons lemon juice
Sour cream for garnish

Don't worry about the wine in this soup intoxicating anybody because the alcohol is cooked off. This recipe differs from the other soup recipes in that the ingredients are cooked and are not pureed. I adapted it from a recipe by Yvonne Young Tarr.

Combine the strawberries, wine, water, and sugar in a saucepan. Bring to a boil over medium heat, cooking and stirring until the sugar is dissolved. Soften the arrowroot in the ¼ cup water and stir it gradually into the soup. Continue cooking and stirring until the soup is thickened and clear. Chill. At serving time, stir in the lemon juice. Serve with a small spoonful of sour cream on top of each bowl.

Cold Strawberry Bisque

SUITABLE FOR SLICED AND
 SWEETENED FROZEN
 STRAWBERRIES
YIELD: 6 SERVINGS

2 10-ounce packages sweetened
 frozen strawberries, thawed
1 cup milk
2 tablespoons instant nonfat dry
 milk
½ cup fresh yogurt (preferably
 homemade)
Flaked coconut for garnish

A recipe for this bisque in a cookbook by the Junior League of Savannah uses cream and sour cream. I cut down on the fats, with equally pleasing results.

Puree all ingredients in a blender or food processor. Chill at least an hour, then stir lightly before serving. Sprinkle a spoonful of coconut over each serving for garnish.

Salads

I fell in with an obliging man who conducted me to the best locality for (wild) strawberries. He said that he would not have shown me the place if he had not seen that I was a stranger, and could not anticipate him another year; I therefore feel bound in honor not to reveal it.

Henry David Thoreau
CAPE COD

Strawberry Fruit Mold

YIELD: 6–8 SERVINGS

1 envelope unflavored gelatin
¼ cup cold water
1½ cups pineapple juice
½ cup orange juice
2 teaspoons mixed pickling spices
1 tablespoon sugar
1 tablespoon lemon juice
1 cup pineapple chunks canned in
    their own juice, drained
2 cups sliced fresh strawberries
¼ cup white grapes
Fresh mint leaves, grapes, and
    strawberries for garnish
Lettuce

A great advantage to molded salads using strawberries is that you can prepare them ahead of time without losing any of the quality of the strawberries. Their juices and color are retained by the gelatin.

Soften the gelatin in the cold water. In a saucepan combine the pineapple and orange juices, pickling spices, and sugar. Bring just to a boil and simmer 2 minutes. Remove the spices. (The easiest way to do this is to tie them in a small cheesecloth bag before putting them into the juice to simmer.) Add the softened gelatin to the hot juice, stirring until dissolved. Add the lemon juice. Chill until the gelatin is about half set, then fold in the pineapple, strawberries, and grapes. Pour into an oiled 4-cup mold. Chill at least 3 hours. To serve, unmold onto a serving plate covered with lettuce leaves and garnish with fresh mint leaves, a few grapes, and whole strawberries.

Southern Strawberry Salad

Citrus and strawberries make a refreshing combination for a summer salad from your garden. I probably don't have to tell you this, but when you're dealing with fresh salad greens from the garden, it's important to give them an extra washing to make sure every bit of grit is gone.

3 cups spinach leaves, torn in bite-sized pieces
3 cups romaine, torn in bite-sized pieces
3 cups watercress, with large stems removed
1 grapefruit, peeled and sectioned
2 oranges, peeled and sectioned
3 cups whole fresh strawberries, washed and hulled
½ cup chopped celery
½ cup chopped green pepper
¼ cup almonds, slivered and toasted

---

1 tablespoon fresh chopped oregano (or 1 teaspoon dried)
1 tablespoon fresh chopped basil (or 1 teaspoon dried)
2 tablespoons fresh chopped parsley (or 1 tablespoon dried)
1 teaspoon salt
1 clove garlic, peeled
⅔ cup salad oil
½ cup grapefruit juice
2 teaspoons grated grapefruit rind
2 tablespoons strawberry puree

Toss the salad greens lightly in a large bowl. Arrange the grapefruit, oranges, strawberries, celery, and green peppers in an attractive pattern on top of the greens. Sprinkle on the almonds.

To make the dressing, combine all the herbs, salt, garlic, salad oil, grapefruit juice, and rind, and strawberry puree in a shaker jar. Chill. At serving time, shake again, remove the garlic, and pour the dressing sparingly over the salad. Pass the remaining dressing in a pitcher with a spoon for diners to help themselves to more.

Strawberry-Avocado Salad

3 cups spinach leaves, torn in bite-sized pieces
1 cup green or red seedless grapes
2 oranges, peeled and sectioned
2 cups whole fresh strawberries, washed and hulled
2 avocados, seeded, peeled, and sliced
¼ cup chopped pecans

1 cup strawberry yogurt (p. 80)
1 teaspoon fresh dill weed (or ½ dried)

Place the spinach leaves in the bottom of a large salad bowl. Arrange the fruit and strawberries on top. Sprinkle with pecans.

To make the dressing, stir the dill into the yogurt and serve in a small bowl for diners to help themselves.

Strawberry-Banana Salad with Sesame

3 medium bananas
1 teaspoon lemon juice
3 cups whole fresh strawberries

⅓ cup mayonnaise
2 tablespoons sour cream
2 tablespoons milk

Lettuce leaves
¼ cup toasted sesame seeds

Although this recipe does not use a lot of sesame seeds, they add a distinctive flavor and an important texture, so don't skip them. You can toast your own by shaking them briefly in a hot skillet until they change color.

Peel and slice the bananas. Sprinkle them with the lemon juice. Wash and hull the strawberries and mix them with the bananas. Make the dressing by stirring together the mayonnaise, sour cream, and milk. Pour over the bananas and berries and mix carefully. Arrange the salad on a bed of lettuce leaves and sprinkle liberally with the sesame seeds. Plan to serve the salad shortly after you mix it so the bananas won't get dark and mushy.

Boston Fruit Salad

YIELD: 6 SERVINGS

6 small heads Boston or bibb
    lettuce
3 cups whole fresh strawberries
2 cups seedless white grapes
1 cup pecan meats, coarsely
    chopped

2 tablespoons honey
2 tablespoons lemon juice
¼ cup mayonnaise
½ cup whipping cream

Wash each head of lettuce, wrap them in a towel and refrigerate for at least 1 hour.

Wash and hull the strawberries. Drain them on a towel. Cut each strawberry in half. Pull the stems from the grapes and wash and drain the grapes. Mix them in a large bowl with the strawberries and pecan meats.

Mix together the honey, lemon juice, and mayonnaise and pour over the fruit mixture. Whip the cream until it holds soft round peaks. Fold it into the fruit salad.

Pull each head of lettuce open to make a leafy bowl. Pile the salad into the lettuce, making six individual servings, and serve immediately.

Romaine-Strawberry Salad

YIELD: 6 SERVINGS

1 large or two small heads of
    romaine lettuce
2 cups whole strawberries

¼ cup salad oil
2 tablespoons lemon juice
1 tablespoon chopped fresh dill (or
    2 teaspoons dried)
¼ teaspoon salt

⅓ cup raw sunflower seeds

To my taste, romaine is the best of all the salad greens. Its piquant taste and the little ridges in its firm leaves contrast perfectly with the slight grittiness and sweetness of strawberries.

Pull off the tough outer leaves of the romaine and discard. Wash the rest of the leaves thoroughly, tear them into bite-sized pieces, and wrap in a towel to chill in the refrigerator.

Wash and hull the strawberries and drain on a towel.

To make the dressing, beat together the oil, lemon, dill, and salt with a fork, or mix in a blender.

Toss the dressing and the romaine together, then add the strawberries and toss again, lightly. Sprinkle the sunflower seeds over the top of the salad and serve at once, on cold plates or in glass bowls.

Spinach Salad with Strawberries and Oranges

YIELD: 6 SERVINGS

1 pound spinach leaves, washed,
   with stems removed
3 green onions (scallions), chopped
2 cups whole strawberries
1 small can (about ¾ cup)
   mandarin oranges, drained
¼ cup cashews, coarsely chopped

3 tablespoons honey
½ teaspoon salt
½ teaspoon dry mustard
½ teaspoon paprika
½ cup salad oil
2 tablespoons vinegar
½ teaspoon celery seed

Here's a perfect accompaniment to seafood or chicken dishes.

Be sure the spinach is very well washed, then dry it, wrap in a towel, and refrigerate to make it crisp. Wash, hull, and drain the strawberries. In a large bowl combine the spinach, chopped onion, strawberries, oranges, and cashews.

To make the dressing, beat all remaining ingredients together with a fork or mix in a blender. Do not pour the dressing over the salad until ready to serve or the spinach will be limp.

Strawberry-Shrimp Salad

YIELD: 6 SERVINGS

1½ pounds cooked, shelled, and
  deveined shrimp
¾ cup diced celery
4 hard-cooked eggs
2 tablespoons chopped green
  pepper

1 teaspoon dry mustard
2 tablespoons chopped fresh
  parsley (or 1 tablespoon dried)
1 tablespoon lemon juice
¾ cup mayonnaise
¼ cup sour cream

2 cups fresh whole strawberries
½ cup cashews

Bibb or Boston lettuce

This is a special dish. Although you can make it successfully with frozen shrimp, the unique sweetness of fresh, unfrozen shrimp complements the sweet-tart flavor of strawberries perfectly. It's really worth the effort to try to find fresh shrimp.

Cut the shrimp into bite-sized pieces if they are large. Leave them whole if they are small. Reserve a few for garnish. Chop the eggs. Mix the shrimp, celery, eggs, and green pepper in a large bowl.

Make the dressing by combining the mustard, parsley, lemon juice, mayonnaise, and sour cream and beating with a fork or processing in the blender. Pour the dressing over the shrimp mixture. Toss lightly until everything is coated.

Add the cashews and strawberries and mix again, very gently. Reserve a few of the strawberries for garnish. Chill about 30 minutes before serving.

To serve dramatically, allow a small head of lettuce for each serving. Pull it open to form a bowl, and fill it with the salad. Or, arrange lettuce leaves on a large serving plate and pile the salad in the middle. Garnish with the reserved shrimp and strawberries.

Strawberry-Chicken Salad in Chiffon Dressing

YIELD: 6 SERVINGS

I assembled this recipe from several sources out of my past. My Grandmother Pennington used to make a special slaw at holiday time that was so delicate and good my father called it "Ice Cream Cabbage." And a restaurant where I once worked made a special chicken salad containing shredded cabbage that was so popular it usually sold out before lunch was over. I added the strawberries for color and texture contrast, and the resulting salad combines a main meal and a dessert in one mouthwatering concoction. The cream for the dressing has to be whipped close to serving time, but you can have all the other ingredients ready so that the last-minute preparations won't take long.

½ medium head of cabbage, finely shredded
½ teaspoon salt
2 cups cooked, diced chicken (preferably white meat)

½ pint whipping cream
3 tablespoons sugar (more or less according to taste)
2 tablespoons vinegar (more or less according to taste)

---

2 cups fresh sliced strawberries
Romaine lettuce

In a large mixing bowl sprinkle the cabbage with the salt and mix thoroughly. Mix in the chicken. Make the dressing by whipping the cream until it holds soft round peaks, then beat in the sugar and vinegar to taste. With a fork stir the dressing into the cabbage-chicken mixture a little at a time, being careful not to drive out the air you have whipped into the cream. When all the dressing has been incorporated, quickly stir in the sliced strawberries. Serve at once on the lettuce leaves.

Cold Salmon with Strawberries in Cucumber Dressing

YIELD: 6 SERVINGS

The combination of colors in this dish is as appealing as the combination of flavors—the pink of the salmon and the red strawberries set off by the deep green of the watercress and delicately finished with the paler green of the cucumber dressing. For a party you can poach the salmon and make the cucumber dressing ahead of time so that all you have to do at the last minute is assemble everything on the serving platter.

1 cup water
1 cup dry white wine
1 onion, sliced
1 celery rib, chopped
6 fresh salmon steaks
2 cups fresh whole strawberries

1 large cucumber, peeled and
   seeded
1 cup sour cream
2 tablespoons chopped fresh
   parsley (or 1 tablespoon dried)
1 tablespoon chopped fresh dill (or
   2 teaspoons dried)
½ teaspoon salt
1 teaspoon lemon juice

Watercress
Fresh chopped chives or green
   onion for garnish

Bring the water, wine, onion, and celery to a boil in a large skillet. Reduce heat to simmer, and poach the salmon steaks, covered, in the liquid for about 10 minutes, or until the salmon is just cooked through. Do not overcook. Cool the salmon in the liquid, then remove any skin from the steaks and chill them.

Wash and hull the strawberries and drain them on a towel.

Make the dressing by combining the cucumber, sour cream, parsley, dill and salt and lemon juice in the blender or food processor. Process until smooth.

To serve, arrange the watercress around the salmon steaks on a serving platter, scatter the strawberries over the salmon, and pour the dressing over all. Garnish with a spoonful of chopped chives or green onion and serve at once.

Strawberry-Salmon Mousse

YIELD: 6 SERVINGS

The combination of strawberries, salmon, and cucumbers may sound unlikely, but it works. When you can't get fresh salmon, this mousse makes good use of canned salmon. The flavor of the strawberries and the crispness of the cucumber give the mousse a fresh taste.

1 package unflavored gelatin
¼ cup cold water
¼ cup boiling water
1 14½-ounce can salmon
½ cup chopped celery
¾ cup sour cream
1 tablespoon lemon juice
½ teaspoon salt
2 tablespoons finely minced onion
1 cup sliced fresh strawberries
½ cup peeled, sliced cucumber
Lettuce

Soften the gelatin in the cold water. Add the boiling water and stir to dissolve. Let the mixture cool. Drain and flake the salmon, removing skin and bones. Add the chopped celery. Stir the gelatin, sour cream, lemon juice, salt, and onion into the salmon mixture. Gently fold the strawberries and cucumber into the salmon mixture.

Oil a 3-cup mold. Pour the mousse mixture into the mold and chill at least 3 hours before serving.

To serve, unmold onto a cold serving plate covered with lettuce leaves. Garnish with whole strawberries and a few cucumber slices. If you want a special dressing, try the combination below.

Dressing for Strawberry-Salmon Mousse

YIELD: 1 CUP DRESSING

½ cup sour cream
½ cup strawberry puree
1 tablespoon lemon juice
1 tablespoon honey

Combine all ingredients and beat with a fork or mix in a blender. Chill at least 30 minutes before serving to give the flavors a chance to blend.

Turkey Fruit Salad

YIELD: 6 SERVINGS

Someday I'm going to write a book called What to Do with the Rest of the Turkey. *This recipe is going to be in it. Think of it as a sneak preview. Plan to serve this salad shortly after you make it so the bananas and apples don't get dark and the strawberries don't get mushy.*

1 tablespoon lemon juice
1 large banana, sliced
1 cup unpeeled apple, cut into
  bite-sized pieces
2 cups cut-up cooked turkey
  (preferably white meat)
2 cups pineapple chunks canned in
  their own juice, drained
3 cups fresh strawberries
½ cup chopped celery
¼ cup coarsely chopped walnuts
2 tablespoons raw sunflower seeds

¾ cup salad oil
¼ cup lemon juice
2 tablespoons pineapple juice from
  can
2 tablespoons honey
2 tablespoons catsup
½ teaspoon salt

Lettuce leaves
Strawberries for garnish

Sprinkle the lemon juice over the banana and apple pieces in a large bowl. Add the turkey, drained pineapple (reserving 2 tablespoons of the juice for the dressing), strawberries, celery, nuts, and sunflower seeds.

To make the dressing, mix together the oil, lemon juice, pineapple juice, honey, catsup, and salt in a blender or shaker jar. Pour over the salad mixture.

Serve on lettuce leaves garnished with a few fresh strawberries. Strawberry bread (page 43) would make a nice accompaniment to this salad.

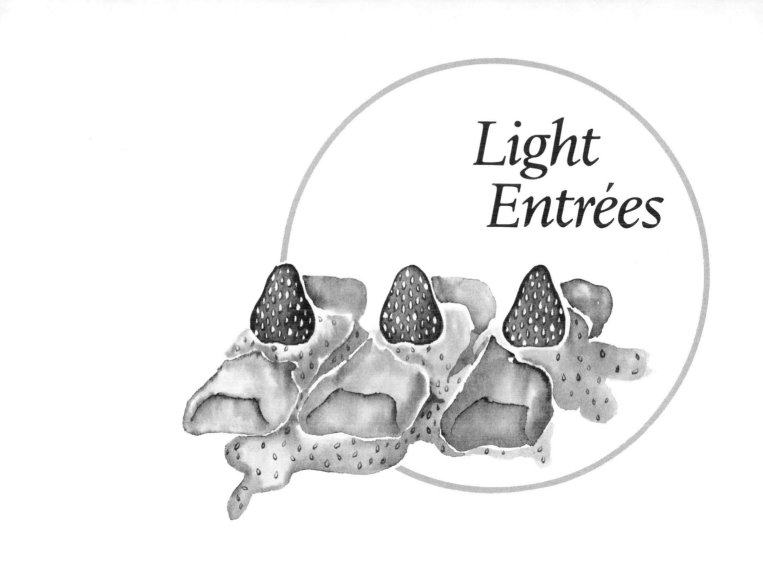

*Light
Entrées*

No one… has really experienced paradise on earth until he has plucked and eaten a clutch of tiny, fully ripe wild strawberries, warmed by mountain sunshine.

Irma S. Rombauer and
Marion Rombauer Becker
JOY OF COOKING

Strawberry Pancakes

SUITABLE FOR USING
  UNSWEETENED FROZEN
  STRAWBERRIES
YIELD: 6 SERVINGS

3 eggs
1 cup unsweetened strawberry
  puree
1¼ cup unbleached white flour
½ cup whole wheat flour
2 tablespoons sugar
¼ teaspoon salt
2½ teaspoons baking powder
2 cups milk
4 tablespoons melted butter
1 tablespoon ground cinnamon

I've never been much of a pancake eater but these are really good—even to me. If you use a one-pound package of commercially frozen strawberries, it will produce about 2 cups of puree: 1 cup for the pancakes and the other cup to spread over them instead of syrup.

Beat the eggs until light. Beat in the strawberry puree. Sift together the flours, sugar, salt, and baking powder. Beat into the puree alternately with the milk, ending with dry ingredients. Stir in the melted butter and the cinnamon. Cook on a hot griddle, (300°F.) turning only once when bubbles cover the unbaked side and the pancake edges begin to look dry.

These pancakes are especially good served with strawberry puree lightly flavored with cinnamon and sprinkled with a small amount of granulated sugar.

Strawberry-Ricotta Custard

SUITABLE FOR UNSWEETENED
  FROZEN STRAWBERRIES
YIELD: 6 SERVINGS

2 cups ricotta cheese
¼ cup milk
4 eggs
¼ cup mild honey
1½ cups fresh or frozen
   strawberries
2 tablespoons cornstarch
½ teaspoon vanilla extract
½ cup fresh strawberries for
   garnish (optional)

Delicious for breakfast or lunch, this has enough protein to keep you going for hours. My kids always thought of it as cheesecake and felt guilty for enjoying it so much, so I suppose you could get away with serving it for dessert.

Preheat the oven to 350°F.

In the blender or food processor beat together the ricotta, milk, eggs, and honey. Mash a few of the strawberries together with the cornstarch until it dissolves. Put the cornstarch mixture and the rest of the 1½ cups strawberries into the food processor with the cheese. Blend until smooth. Blend in the vanilla.

Butter a 6-cup baking dish and dust it with granulated sugar. Pour the strawberry-cheese mixture into the dish and place in a shallow pan of hot water. Bake in a preheated oven about 1 hour, or until the custard is barely baked through. A silver knife inserted near the center should come out *nearly* clean. The center may still be slightly soft but it will set as it cools. Overbaking will make the custard coarse and watery.

Remove the dish from the hot water bath and cool to room temperature before serving. It is good chilled, too. If you are using the strawberry garnish, mix the remaining berries in the blender or food processor and pour over the wedges of custard like a sauce.

Fried Cheese with Strawberry Sauce

YIELD: 6 SERVINGS

At Hennessey's restaurant in Columbia, South Carolina, fried cheese with strawberry sauce is a featured item on the appetizer menu, but it's substantial enough a dish to enjoy for brunch or lunch. My version is somewhat simpler than Hennessey's. I think it tastes every bit as good.

You can buy a couple of small Gouda rounds to make the fried cheese, or have the slices cut from a large cheese in the delicatessen. Monterey Jack cheese is an adequate substitute for the Gouda, but the flavor will be blander.

Fried Cheese

12 ½-inch thick slices of Gouda
  cheese
Flour
1 egg
Dry bread crumbs
Oil for frying

Dredge the cheese slices in the flour. Beat the egg. Dip the cheese in the beaten egg. Roll it in bread crumbs. Allow the crumbed cheese slices to dry several minutes before frying.

To fry, pour enough oil in a heavy skillet to cover the bottom by about ¼ inch. Heat to medium high and fry the cheese first on one side, then the other, turning only once. The slices should be crisp and brown on the outside and soft inside. Serve immediately, topped with the chilled strawberry sauce below.

Strawberry Sauce for Cheese

1 cup water
1 cup sugar
2 ½ cups sliced fresh strawberries
2 tablespoons cornstarch
2 tablespoons lemon juice

Combine the water and sugar in a saucepan and bring to a boil. Add the strawberries. Lower the heat and simmer 3 minutes. Dissolve the cornstarch in the lemon juice and stir it into the sauce. Cook and stir about 3 minutes more, or just until the sauce becomes clear and slightly thick. Cool, then chill in the refrigerator. The sauce will keep several weeks in the refrigerator.

Strawberry Blintzes

Here are several versions of strawberry blintzes—with the strawberries on the inside, with the strawberries on the outside, and one with a strawberry-cheese filling. You can use one of the suggested strawberry sauces or strawberry puree as a topping. No matter how you fill it or sauce it, the blintz shell, basically just a thin pancake, remains the same. If you've never made blintzes, you may have trouble with the first couple, but you'll get the knack quickly, so keep trying.

Blintz Batter

3 eggs
1 cup milk
2 tablespoons vegetable oil
¾ cup unbleached white flour, sifted

Beat all the ingredients together in a blender or with a mixer, adding the flour a little at a time to avoid lumps.

Rub the bottom of a heavy 6-inch skillet (iron is best) generously with oil. Heat it until a drop of water sizzles on the surface. Pour in about 2 tablespoons of the blintz batter and tilt the pan so the batter makes a thin covering across the bottom. Set it back onto medium-high heat and let it cook until the bottom is brown and the blintz is done through—two to three minutes. Turn it out onto a towel and continue cooking the blintzes until all batter is used. When you become practiced at this you can easily keep two 6-inch skillets going, or you can use a crepe maker if you wish.

Spread a heaping tablespoon of filling along one side of the blintz shell, fold in the ends, and roll the blintz like an egg roll.

Brush the blintzes lightly with oil and bake in a 425°F. oven about 15 minutes, or until browned and heated through. Alternatively, cover the bottom of a large skillet to ¼ inch with equal amounts of melted butter and oil, heat, and brown the blintzes on both sides.

The traditional serving of blintzes is three on a plate.

Strawberry Filling for Blintzes

YIELD: FILLING FOR 18 BLINTZES

2 cups sliced fresh strawberries
3 tablespoons sugar
1 rounded tablespoon cornstarch
¼ teaspoon cinnamon

Toss all ingredients together and put a small spoonful into each shell. Bake or fry as directed on page 38. Top with sour cream and strawberry puree.

Cheese Filling for Blintzes with Strawberry Sauce

YIELD: ENOUGH FILLING FOR 18
BLINTZES

2 cups dry or small-curd cottage
cheese
1 egg
½ teaspoon salt
1 tablespoon melted butter
2 tablespoons honey
2 teaspoons lemon juice

Beat all ingredients together and put a small spoonful into each blintz shell. Bake or fry as directed on page 38. Serve with strawberry sauce.

Strawberry Cheese Filling for Blintzes

YIELD: ENOUGH FILLING FOR 18
    BLINTZES

1½ cups dry or small-curd cottage
    cheese
1 egg
1 tablespoon melted butter
1 teaspoon cornstarch
1 tablespoon brown sugar
1 cup sliced fresh strawberries

Mix together the cheese, egg, butter, cornstarch and brown sugar until smooth. Gently stir in the sliced strawberries, pouring off any juice that may have accumulated. Place a large spoonful in each blintz shell. Bake or fry according to directions in the basic recipe. Serve with the traditional garnish for cheese blintzes, sour cream. Add a few whole strawberries on top, if you wish.

Strawberry Sauce for Blintzes

SUITABLE FOR USING SLICED AND
    SWEETENED FROZEN
    STRAWBERRIES
YIELD: SAUCE FOR 18 BLINTZES

1 cup sliced fresh strawberries
    sweetened to taste *or* 1 10-ounce
    package frozen sliced
    strawberries, thawed
2 teaspoons cornstarch
¼ cup dry white wine

If using fresh strawberries, sweeten them and allow them to stand for at least 1 hour before proceeding.

Bring the strawberries and their juice to a boil over medium heat. Dissolve the cornstarch in the wine and gradually stir into the simmering strawberries. Cook and stir over medium heat about 5 minutes, or until the sauce is clear and thickened. If the sauce seems too thick add a little more wine or water. Serve warm over cheese-filled blintzes.

Cold Strawberry Sauce

YIELD: 1½ cups

1 cup unsweetened strawberry
  puree
⅓ cup granulated sugar
¼ cup water

Use this sauce for ice cream, plain custard, or yogurt, as well as pancakes, blintzes, and French toast. It will taste better if you make it when you have fresh berries and freeze it than if you try to make it later from frozen and thawed berries.

Combine all ingredients in a saucepan, bring to a boil, and cook for 1 minute. Chill.

Strawberry Butter

SUITABLE FOR USING
  UNSWEETENED FROZEN
  STRAWBERRIES
YIELD: ABOUT 4 CUPS

1 10-ounce package frozen
  strawberries, thawed
1 cup soft butter
2 cups confectioners' sugar

Use strawberry butter on pancakes and waffles instead of syrup. For a wonderfully intense combination of strawberry flavors, try strawberry butter on strawberry pancakes, topped with a few fresh strawberries!

Combine all ingredients in the bowl of a food processor and mix until smooth. Pack into an airtight container and refrigerate. The butter will also keep several weeks in the freezer. You can make this with a blender rather than a food processor, if you do it in two batches.

Strawberry-Honey Butter

YIELD: ABOUT 4 CUPS

3½ cups whole fresh strawberries
½ cup honey
2 tablespoons lemon juice
¾ cup butter

Strawberry pancakes and crepes taste great with the enhancement of honey in the strawberry butter.

Puree the strawberries in the blender or food processor. Pour them into a saucepan with the honey, bring to a boil, and simmer over low heat 20 minutes, stirring from time to time. Stir in the lemon juice and chill the mixture. Have the butter at room temperature. Blend it into the strawberry-honey mixture with a blender, food processor, or electric mixer. Store in the refrigerator.

Strawberry Cream Cheese Spread

YIELD: ABOUT 2 CUPS SPREAD

1 8-ounce package cream cheese
1 cup sliced fresh strawberries or
⅔ cup unsweetened strawberry
   puree

Beat together the cream cheese and strawberries until the mixture is creamy. (A food processor is the easiest way.) Chill overnight to develop flavor. Use for sandwich filling.

Variation: **Strawberry Nut Spread.** Add 2 tablespoons milk and ½ cup finely chopped nuts to make a strawberry-nut spread.

Strawberry-Nut Bread

YIELD: 2 LARGE or 3 SMALL LOAVES

Serve strawberry-nut bread plain in thin slices to accompany a salad, or spread it with Strawberry Cream Cheese Spread (page 42) to make sandwiches for lunch boxes or snacks. The recipes for the bread and the cheese spread came to me from the South Carolina Department of Agriculture, which actively promotes strawberries in the state all spring.

3 cups all-purpose, unbleached
   white flour
1 teaspoon baking soda
½ teaspoon salt
3 teaspoons cinnamon
2 cups sugar
2 cups sliced fresh strawberries
4 eggs
1¼ cups oil
1 cup chopped nuts

Sift the dry ingredients together into a large mixing bowl. Make a well in the center. Beat together the strawberries, eggs, and oil and pour into the well. Stir just enough to dampen all ingredients. Quickly stir in the nuts. Pour into two or three well-greased bread pans, filling no more than half full. Bake in a preheated 350°F. oven 45 to 60 minutes or until the bread is baked through and a toothpick inserted near the center comes out clean. Cool 20 to 30 minutes before removing from pans and finish cooling on a rack. Do not slice until the bread is completely cold.

Desserts

In Napoleonic times, Madame Talien had 22 pounds of strawberries crushed every time she had a bath so that her skin would be soft and smooth. The strawberries in our markets are often pulpy enough to use as a skin freshener such as she would have enjoyed...

Moira Hodgson
THE GOURMET SHOPPER

Strawberry Compote

YIELD: 6 SERVINGS

6 cups fresh whole strawberries
¼ cup honey
1 tablespoon grated orange rind
2 tablespoons orange juice
1 tablespoon lemon juice

The combination of flavors in this compote has a naturalness that calls up visions of simple living along the river or pre-apple days in the Garden of Eden. Be sure to use a mild-flavored honey.

Wash and hull the strawberries. Drain them on a towel. Combine the honey, orange rind, and juices in a large bowl. Add the strawberries and stir gently to coat them with the honey mixture. Allow them to stand about 30 minutes at room temperature to draw juice, then refrigerate until serving time. Don't keep the strawberries refrigerated for much more than an hour.

Strawberries Grand Marnier

YIELD: 6 SERVINGS

4 cups sliced strawberries
¼ to ⅓ cup sugar, according to taste
1 tablespoon grated orange rind
¼ cup Grand Marnier

This version of strawberries and Grand Marnier liqueur has some added sugar and orange flavors. Use it to enhance berries that are not quite perfectly sweet.

Combine all ingredients and chill at least 1 hour before serving.

Strawberries with Champagne Jelly

YIELD: 8 SERVINGS

You might find it a bit of a nuisance getting the raspberry syrup for this recipe and you could substitute ¼ cup lemon juice and ¼ cup granulated sugar, but I think it's worth the bother to have that nice raspberry-strawberry flavor combination.

2 envelopes unflavored gelatin
1 cup cold water
¼ cup syrup from frozen red
   raspberries
3 cups champagne

4 cups sliced fresh strawberries
¼ cup brown sugar

To make the jelly, soften the gelatin in the cold water, then heat over low heat and stir until the gelatin dissolves. Stir in the raspberry syrup and the champagne. Chill in the refrigerator until firmly set, at least 3 hours.

Sprinkle the strawberries with the brown sugar and allow to stand at room temperature about 30 minutes. Stir to dissolve the sugar.

To serve, break up the jelly into small pieces with a fork until it is quite spongy. Spoon into individual serving dishes and top with strawberries.

Strawberries in Cranberry Jelly

YIELD: 4–6 SERVINGS

The only thing I like about commercial strawberry gelatin is its bright red color. Cranberry juice cocktail provides the nice color in this recipe.

1 envelope unflavored gelatin
2 cups cranberry juice cocktail
2 cups whole fresh strawberries,
   washed and hulled
2 apples, cored and chopped (skins
   left on)

Soften the gelatin in the cranberry juice cocktail, then heat and stir until the gelatin dissolves. Chill until the gelatin mixture is about half set. Stir in the strawberries and chopped apples and chill until firm. Serve as a simple dessert or what southerners call a "congealed" salad.

Strawberry Ricotta

YIELD: 6 SERVINGS

1 pound whole-milk ricotta cheese
3 tablespoons honey
½ teaspoon vanilla
2½ cups sliced fresh strawberries

Bring the ricotta to room temperature. Beat it together with the honey and vanilla until smooth. Fold in the strawberries and spoon the mixture into 6 small dishes. Serve for breakfast or as a light dessert.

Homemade Strawberry Yogurt

YIELD: SLIGHTLY OVER 1 QUART

I still remember the first time I tried this. The yogurt set with such perfection and tasted so fine it seemed like magic. Nothing in the dairy case at the grocery store comes close to homemade yogurt for good taste.

1 quart whole milk
1 tablespoon yogurt (for starter)
1 cup fresh unsweetened
   strawberry puree
1 tablespoon granulated sugar
   (optional or to taste)

Scald the milk and cool to lukewarm (110°F.) Have the pureed strawberries and the starter at room temperature. Stir them into the cooled milk. Add sugar if you are using it and stir until it dissolves. Pour the mixture into the container of your yogurt maker or into warmed wide-mouth thermos bottles. Incubate until the mixture is set, 4 to 8 hours. Refrigerate immediately.

Because of the fresh fruit in it, this yogurt will begin to ferment and get fizzy in about 3 days. If you can't eat a quart of yogurt that fast, divide the recipe in half.

This yogurt is especially good served with lots more fresh, sliced strawberries on top.

Baked Strawberry Custard

SUITABLE FOR USING
  UNSWEETENED FROZEN
  STRAWBERRIES
YIELD: 6 SERVINGS

3 cups milk
4 eggs
¼ cup honey (or to taste)
½ teaspoon vanilla extract
¼ teaspoon ground nutmeg
2 cups whole strawberries, fresh or
  frozen and thawed

The little country restaurant about a mile down the road from my house serves this at dinner (noon) at least once a week. It's so popular that unless I get there early it's apt to be sold out. It's easy to make. The waitresses tell me they do it at home all the time.

Beat together the milk, eggs, honey, vanilla, and nutmeg using a blender, food processor, or rotary beater. Pour the mixture into 6 individual custard cups or one 6-cup baking dish. Put in the strawberries, dividing them equally if you used individual cups.

Put the cups or baking dish into a pan of hot water and bake the custard in a preheated 350°F. oven 20 to 40 minutes, depending on the size of the cups.

Remove the custard from the oven before it is fully set and allow it to cool standing in the pan of hot water. This will finish cooking the custard without its becoming "weepy," although liquid sometimes accumulates around the strawberries, which float on top of the custard.

Strawberries with Yogurt Snow

YIELD: 6 SERVINGS

1 cup plain yogurt (preferably
  homemade)
1 teaspoon lemon juice
1 teaspoon brandy
1 egg white
2 tablespoons sugar
6 cups whole fresh strawberries

My daughter sent me this recipe from England. I converted the ingredients from weights to our more familiar measures

Mix the yogurt, lemon juice, and brandy together. Beat the egg white until it is stiff, then beat in the sugar. Continue beating until the mixture is glossy and forms stiff peaks. Fold in the yogurt. Wash and hull the strawberries. Put them into individual sherbet dishes and top with the yogurt snow. Serve immediately.

Strawberry-Orange Yogurt Dessert

YIELD: 6–8 SERVINGS

Don't let on how easy this is. It will establish your reputation as a good cook. The original version was made with sour cream instead of yogurt, which I found too rich to be enjoyable. Yogurt, especially the sweet, homemade kind, produces an equally tasty and certainly more healthful dessert.

4 cups whole fresh strawberries
½ cup orange juice
⅔ cup brown sugar
2 cups plain yogurt (preferably homemade)

Wash and hull the strawberries and drain them on a towel. Marinate them in the orange juice for an hour. Combine the brown sugar and yogurt. Stir in the strawberries and the juice in which they were marinating. Serve in tall-stemmed sherbet glasses.

Strawberries Romanoff

YIELD: 6 SERVINGS

Definitely not original with me, all versions of this classic party dessert are about the same. If you'd like a little variety you could substitute strawberry ice cream for the vanilla in this recipe. The Blender Ice Cream on page 76 would be fine.

4 cups whole fresh strawberries
1 cup whipping cream
1 pint (2 cups) vanilla ice cream, softened
2 tablespoons Kirsch

Wash and hull the strawberries. Drain them on a towel and put them into a large mixing bowl. Whip the cream until it forms soft peaks. Soften the ice cream. Fold the whipped cream and Kirsch into it. Pour the ice cream mixture over the strawberries and refrigerate until serving time, up to one hour.

To serve, spoon into individual serving dishes, stirring as you do so to keep the strawberries mixed into the sauce.

Strawberry-Avocados Romanoff

SUITABLE FOR UNSWEETENED
 FROZEN STRAWBERRIES
YIELD: 4–6 SERVINGS

2 avocados, seeded, peeled, and diced
1 tablespoon lemon juice
12 large fresh or frozen strawberries (partly thawed if frozen)
¾ cup strawberry yogurt (p. 80)
1 tablespoon orange juice concentrate
1 teaspoon brown sugar
Shredded coconut for garnish

Although developed by the California Avocado Commission to encourage consumers to use more avocados, this recipe does wonderful things for strawberries, too.

Sprinkle the avocados with lemon juice. Mash the strawberries in a bowl with a fork. Gently stir in the diced avocado. Spoon the mixture into 4–6 individual serving dishes.

Combine the strawberry yogurt, orange juice concentrate, and brown sugar. Pour over the strawberry-avocado mixture. Sprinkle with coconut. Serve at once or chill up to an hour before serving.

Grand Champion Strawberry Soufflé

SUITABLE FOR UNSWEETENED
  FROZEN STRAWBERRIES
YIELD: 8–10 SERVINGS

6½ cups whole strawberries
2 envelopes unflavored gelatin
½ cup water
⅔ cup sugar
4 egg yolks, well beaten
⅛ teaspoon salt
1 tablespoon lemon juice
4 egg whites
½ cup sugar
1 cup whipping cream
Whipped cream and a few fresh
  strawberries for garnish
  (optional)

This recipe won an award in a Florida strawberry contest in 1979. It's a good make-ahead dessert.

Wash and hull the strawberries. (If you are using frozen strawberries, thaw just long enough to remove the ice crystals from the center of the berries.) Puree them in the blender or food processor.

In a saucepan, combine 1 cup of the puree, the gelatin, water, sugar, egg yolks, salt, and lemon juice. Stir to soften the gelatin, then heat the mixture just to a boil. Cool, then refrigerate until mixture has the consistency of unbeaten egg whites, about 30 minutes. Meanwhile, beat the egg whites until they form stiff, glossy peaks. Gradually beat in the ½ cup sugar. Whip the cream until it holds soft peaks. Fold the whipped cream into the egg whites, then fold the chilled berry mixture into the cream–egg white combination. Pour into an oiled 8-cup mold and chill at least 3 hours before serving.

To serve, unmold onto a chilled plate and garnish with whole fresh berries and whipped cream.

Molded Strawberry Cream

SUITABLE FOR UNSWEETENED
    FROZEN STRAWBERRIES
YIELD: 12 SERVINGS

1 envelope unflavored gelatin
¼ cup cold water
2 cups unsweetened strawberry
    puree
1 cup sugar (or to taste)
2 cups whipping cream
½ teaspoon vanilla extract (or to
    taste)
Fresh mint leaves and orange slices
    for garnish

Don't think of this as daily fare. It's too rich. But all that wonderful whipped cream tastes divine in this recipe for strawberries and cream with a gelatin base.

Soften the gelatin in the water, then heat until it dissolves. Combine it with the strawberry puree and sugar. Stir until the sugar dissolves. Chill about 20 minutes. Whip the cream until it forms soft peaks. Whip in the vanilla. Fold the whipped cream into the strawberry purée and pour into an oiled 6-cup mold. Chill at least 3 hours before serving. To serve, unmold onto a chilled platter and garnish with fresh mint leaves and thin slices of orange.

Pies and
Cakes

*We are now so conditioned
to cottony strawberries in January that we
have forgotten the pleasures of anticipation, and
the excitement when they finally came
in ripe, tart, and delicious.*

John L. and Karen Hess
THE TASTE OF AMERICA

Fresh Strawberry Pie I

YIELD: 1 9-INCH PIE (6–8 SERVINGS)

You've seen many, many versions of fresh strawberry pie, I'm sure. The trouble with most of them is that the strawberries are glazed either with a commercial glaze or with strawberry-flavored gelatin, both of which contain artificial color and flavor and taste more like chewing gum than fresh strawberries. This recipe uses unflavored gelatin. I have not used food coloring, which means the pie will be less sparkling red than those you see in bakeries and on pastry carts. You can certainly add coloring if you wish. It's had such bad press in recent years I prefer not to mess with it. This is the most basic of the fresh strawberry pies.

4 cups fresh whole strawberries
1 cup sugar
1 envelope unflavored gelatin
¼ cup cold water
½ cup sugar
2 tablespoons lemon juice

1 9-inch baked pie shell

Whipped cream and whole
   strawberries for garnish

Wash and hull the strawberries. Put them in a large bowl and cover with 1 cup sugar. Refrigerate for at least 3 hours to draw juice.

Soften the gelatin in the cold water. Pour the juice and about half the strawberries into a blender or food processor and puree. (Or, if you want a clearer juice, press through a sieve.) You should have 1½ *cups juice and berry puree.* Add water if necessary to make the full amount. Pour into a saucepan, stir in the ½ cup sugar and lemon juice, and heat just to boiling. Add the gelatin and stir until dissolved. (If you choose to use red food coloring, add it now.) Refrigerate about 30 minutes or until the mixture begins to set.

Arrange the remaining whole strawberries in the pie shell and pour the thickening gelatin mixture over them. Chill at least 3 hours before serving.

Garnish with whipped cream and a few more strawberries to serve.

Fresh Strawberry Pie II

6 cups fresh whole strawberries
½ cup granulated sugar
1 tablespoon lemon juice
1 tablespoon cornstarch

1 9-inch baked pie shell

Whipped cream and whole
strawberries for garnish

This recipes differs from the previous one in using cornstarch rather than gelatin to thicken the glaze. Also, it is much less sweet.

Wash and hull the strawberries. Drain half of them on a towel. Puree the rest. (Strain through a sieve if you want a clearer glaze.) Combine the puree in a saucepan with the sugar, lemon juice, and cornstarch. Cook and stir over medium heat until the mixture becomes clear and thick. Cool.

Arrange the reserved strawberries in the pie shell and pour the glaze over all. Chill at least 1 hour before serving. Garnish with whipped cream and strawberries.

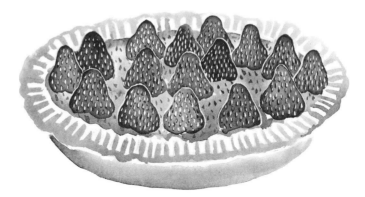

Fresh Strawberry Pie III

YIELD: 1 9-INCH PIE (6–8 SERVINGS)

1 3-ounce package cream cheese
3 tablespoons Cointreau or other orange liqueur
5 cups whole fresh strawberries
½ cup sugar
¼ cup water
1½ tablespoons cornstarch
¼ cup water
1 tablespoon lemon juice

1 9-inch baked pie shell

Whipped cream and whole strawberries for garnish

A thin layer of liqueur-flavored cream cheese on the bottom of the crust distinguishes this fresh strawberry pie from the others.

Bring the cream cheese to room temperature and beat the Cointreau into it. Wash and hull the strawberries. Slice enough to make 1 cupful. Put the sliced berries into a saucepan with the ½ cup sugar and ¼ cup water. Bring to a boil and cook 3 minutes, stirring constantly. Dissolve the cornstarch in the other ¼ cup water and stir it into the simmering berries. Cook and stir until the mixture is thickened and clear. Stir in the lemon juice and remove from heat. Cool.

Spread the cream cheese in the bottom of the pie crust. Arrange the remaining whole berries on top of the cream cheese. Pour the glaze over all. Refrigerate at least 2 hours before serving. Garnish with whipped cream and strawberries.

Baked Strawberry Pie

YIELD: 1 9-INCH PIE (6–8
SERVINGS)

Pastry for a double-crust 9-inch pie

4 heaping cups strawberries
1 cup sugar
2 tablespoons cornstarch
1 tablespoon lemon juice
2 tablespoons butter

When I was in high school I worked at Biggs' American Restaurant, an old-fashioned restaurant where we served chicken and dumplings, homemade rolls, and pot roast. The most popular dessert was a strawberry pie, served warm, with a scoop of vanilla ice cream on top. For years memories of that pie haunted me; I never could track down the Biggs to get the recipe. Here's what I've worked out, as close an approximation as possible.

Preheat the oven to 450°F.

Wash and hull the strawberries and drain them on a towel. Stir together the sugar and cornstarch in a large bowl and mix in the berries. Sprinkle on the lemon juice. Let the berries stand for about 15 minutes.

Line a 9-inch pie pan with pastry. Pour in the strawberry filling, scatter slivers of butter on top, and cover with a top crust. Seal the edges tightly. Prick a few holes in the center of the crust to allow steam to escape, but don't make them too big or the juice will run out.

Bake 10 minutes in a 450°F. oven. Reduce heat to 350°F. and bake 40 to 45 minutes more, or until the crust is golden brown and the filling is thick and bubbling. Brush the top crust with milk once or twice during the last 15 minutes of baking. Remove from the oven and allow to cool at least 30 minutes before serving. The pie is good warm or cold.

Strawberry Chiffon Pie

YIELD: 1 9-INCH PIE (6–8
  SERVINGS)

1 envelope unflavored gelatin
¼ cup cold water
3 egg yolks
½ cup granulated sugar
1 tablespoon lemon juice
1 cup strawberry puree
3 egg whites
¼ cup granulated sugar

1 9-inch baked pie shell

Sweetened whipped cream and
  whole strawberries for garnish

You've doubtless had lemon chiffon pie. Here's the same idea, made with strawberries.

Soften the gelatin in the cold water. In the top of a double boiler beat the egg yolks together with the sugar and lemon juice. Cook and stir over boiling water until the mixture thickens. Add the gelatin and stir until it dissolves. Remove from heat and stir in the strawberry puree. Chill about 30 minutes, or until the mixture begins to set. Beat the egg whites until they hold stiff peaks, then beat in the ¼ cup sugar, a little at a time, and continue beating until the egg whites are stiff and glossy. Fold the egg whites into the strawberry mixture and pour into the prepared pie shell. Chill at least 3 hours before serving. Garnish with whipped cream and strawberries.

Deep Dish Strawberry-Rhubarb Pie

YIELD: 6 SERVINGS

Pastry, enough for a single-crust pie

3 cups strawberries
2 cups rhubarb cut in ½-inch pieces
3 tablespoons cornstarch
¾ cup granulated sugar (or to taste)
2 tablespoons butter
1 teaspoon cinnamon

No collection of strawberry pie recipes would be complete without this classic combination. Because of the high water content in both strawberries and rhubarb, this pie is always very juicy. Be sure to use a deep dish.

Preheat the oven to 375°F.

Combine the strawberries and rhubarb with the sugar and cornstarch, mixing well to coat all the fruit. Put into a deep 6-cup baking dish. Cut the butter into small pieces and scatter them across the strawberry and rhubarb. Sprinkle with cinnamon. Roll out the pastry to form a circle slightly larger than the top of the baking dish. Cut the pastry into strips about ½ inch wide and lay them across the top of the strawberry and rhubarb mixture to form a close lattice-work top.

Bake for about 40 minutes or until the crust is brown and the juice from the rhubarb and strawberries is thick, clear, and bubbling.

Strawberry Alaska Pie

YIELD: 8 SERVINGS

4 cups homemade strawberry ice cream, softened enough to dish
1 10-inch pie shell, baked
2½ cups sliced fresh strawberries

4 egg whites, at room temperature
½ teaspoon cream of tartar
6 tablespoons sugar
½ teaspoon vanilla

Here's another recipe I find too elaborate for daily use but just right for a special occasion. As sophisticated as we've become about cooking, the idea of ice cream baked into a pie has never lost its magic for most of us.

To assemble the pie, spread the ice cream in the pie shell. Top with the strawberries. Freeze until the ice cream is hard again. Shortly before you are ready to serve the pie, prepare the meringue topping.

Meringue Topping

Preheat the oven to 450°F.

To prepare the meringue topping, beat the egg whites until they are foamy. Add the cream of tartar and continue beating until the whites are stiff enough to form soft peaks. Begin beating in the sugar, a spoonful at a time. Beat in the vanilla. The mixture should be quite stiff and glossy, but not dry.

Remove the pie from the freezer. Working from the edges toward the center, swirl the meringue completely over the top of the pie. Make sure it touches the edges of the pie plate all way way around to make a seal. Bake for 5 minutes or until the meringue is lightly browned. Serve immediately.

Variation: Instead of a pie shell, use a cake layer, or thick slices of pound cake, on a baking sheet. Mound the ice cream in the center and freeze until hard. At serving time spread the meringue over all, covering the sides of the cake down to the baking sheet. Bake as above.

Strawberry-Kiwi Tart

1 9-INCH TART (6–8 SERVINGS)

Kiwi fruit has a beautiful clear green color and its dark seeds make an attractive pattern on its slices. Both its color and flavor contrast appealingly with strawberries.

3 egg yolks
¾ cup granulated sugar
Juice of 1 lemon
1 teaspoon grated lemon rind
3 egg whites

About 20 whole fresh strawberries
1 ripe kiwi fruit

1 9-inch baked pie shell

In the top of a double boiler beat together the egg yolks, sugar, lemon juice, and rind until light and foamy. Cook over boiling water, stirring constantly, for about 10 minutes or until the mixture is thick. Cool.

Beat the egg whites until they are stiff. Fold into the cooled pudding mixture. Spread this in the tart shell and chill at least 1 hour before serving.

At serving time, cut the strawberries in half and peel and slice the kiwi fruit in thin circles. Arrange the berries and kiwi slices in alternate concentric circles on top of the tart filling.

Strawberry Cheese Pie

YIELD: 8 SERVINGS

Some people call this a cheese cake but to me anything in a pie plate is a pie. It's a good idea to make your own graham cracker crust rather than buying one, because you can be sure you make it in a plate deep enough to hold all the filling and strawberries. Besides, it'll taste better.

Graham Cracker Crust

15 graham crackers
2 tablespoons butter

Crush the crackers into crumbs. Melt the butter and mix it into the crumbs. Press the mixture onto the sides and bottom of a deep 10-inch pie plate. Chill overnight or bake for 10 minutes in a 350°F. oven and cool.

Cheese Filling

2 eggs
9 ounces (3 small packages) cream
   cheese
½ cup sugar
Juice of ½ lemon
⅛ teaspoon salt

Preheat oven to 350°F.

Mix all ingredients with a mixer or in a blender or food processor until smooth and creamy. Pour into the prepared graham cracker crust and bake for 25 to 30 minutes, or until the pie is just set. Remove from the oven and raise the heat to 450°F. while you prepare the topping.

Topping

1 cup sour cream
2 teaspoons sugar
1 teaspoon vanilla extract

Stir the sour cream, sugar, and vanilla together and spread on top of the pie. Return to the oven and bake for 5 minutes at 450°F. Remove the pie from the oven and cool completely before putting on the strawberry layer.

Strawberry Layer

4 cups sliced strawberries
1 cup sugar
⅓ cup cornstarch
½ teaspoon red food coloring
   (optional)
½ cup sour cream

Mix the sliced strawberries and sugar. Refrigerate at least 3 hours to give juice time to form. Drain the juice into a saucepan. Spoon a small amount of the cold juice into the cornstarch to dissolve it, then stir the dissolved cornstarch back into the juice in the pan. Bring to a boil over medium heat, stirring constantly. Cook and stir for 5 minutes, or until mixture is thick and clear. Remove from heat and cool. Stir in the strawberries and sour cream and spread over the cheese pie. Chill at least 1 hour before serving.

Strawberry Angel Pie

SUITABLE FOR USING
   UNSWEETENED FROZEN
   STRAWBERRIES
YIELD: 1 9-INCH PIE (6–8
SERVINGS)

3 egg whites
⅛ teaspoon salt
¼ teaspoon cream of tartar
¾ cup granulated sugar

2 cups strawberry puree
1 envelope unflavored gelatin
1 cup heavy cream
1 teaspoon vanilla extract
Sugar to taste

Whenever you see the term "angel pie" it means that the base is a hard meringue shell. The filling usually is predominately whipped cream. This version includes gelatin so that you can prepare it ahead rather than being stuck trying to assemble it at the last minute.

Meringue Shell

Preheat the oven to 250°F.

Bring the egg whites to room temperature. Add the salt and beat until foamy. Add the cream of tartar and continue beating until the whites form soft peaks. Begin beating in the sugar, a tablespoonful at a time, and continue until the whites are glossy and form stiff peaks.

Spread the meringue on a 9-inch pie plate, building up the edges to form a shell. Bake for about 40 minutes, or until the meringue is hard on the outside and still slightly tender inside. (If you want to store the meringue rather than using it immediately, turn off the oven and leave the meringue shell in it to dry out completely.)

Remove the meringue from the pan to a serving platter before it cools. Cool completely before filling.

Strawberry Angel Filling

Place the puree in a saucepan and soften the gelatin in it. Stir over medium heat until the gelatin dissolves. Refrigerate until the puree begins to thicken and set.

Whip the cream until it holds soft peaks. Beat in the vanilla and sugar to taste. Don't add too much sugar because the meringue shell will be very sweet. Fold the cream into puree and pour into the meringue shell. Chill at least 3 hours before serving.

Variation: Use sliced strawberries instead of puree. Eliminate the gelatin and fold the berries into the sweetened, flavored whipped cream just before serving. No refrigeration will be needed.

Strawberry Meringue Tart

YIELD: 1 9-INCH TART (6–8 SERVINGS)

1 9-inch meringue shell (page 66)
1 8-ounce package cream cheese
3 tablespoons Cointreau or orange juice
1 cup strawberry preserves or jam (page 89)

*The inspiration for this recipe came from **Savannah Style**, a cookbook by the Junior League of Savannah, Georgia. The original recipe was for tiny individual tarts and used fresh strawberries. My version uses strawberry preserves, on the theory that it's easy to find ways to use fresh strawberries and that in mid-winter, a new way to use strawberry preserves might be very appealing.*

Soften the cream cheese at room temperature and beat the Cointreau or orange juice into it to make it easy to spread. Use a little more than called for if the cheese seems too stiff. Spread the softened cheese on the bottom of a cooled meringue shell. Cover with the preserves or jam. Chill about 30 minutes before serving.

Strawberry Shortcake

These days all kinds of concoctions are sold as "strawberry shortcake." While they all have fans, they're not really shortcake. The authentic shortcake recipes are more like biscuits than cake and they're not very sweet. Here's a rich favorite of mine.

2 cups all-purpose, unbleached
    white flour
1 tablespoon baking powder
1 tablespoon granulated sugar
½ teaspoon salt
¼ cup butter
2 eggs
⅓ cup light cream

Preheat the oven to 425°F.

Sift together the flour, baking powder, sugar, and salt. Cut in the butter using a pastry blender or two knives until it is like small peas in the flour. Make a well in the center. Beat the eggs. Combine the beaten eggs and cream and pour into the well in the flour. Mix the dough lightly, as you would for biscuits, just long enough to make the dough hold together. On a lightly floured surface, pat the dough flat to about ½ inch thick. Use a biscuit cutter or large inverted glass to cut 6 individual cakes. (Gather up the scraps to form an extra cake for somebody who wants seconds.) Place the cakes on a lightly greased cookie sheet and bake in a preheated 425°F. oven 10 to 15 minutes, or until brown.

When the cakes are done, split them while they are hot, spread with butter, and fill and cover with strawberry filling.

This kind of shortcake is traditionally served with heavy cream passed in a pitcher to pour over each serving.

Strawberry Shortcake Filling

7½ cups sliced strawberries
½ cup granulated sugar
1 cup whole strawberries

Mash the sliced strawberries lightly with a fork as you stir in the sugar. Refrigerate for at least 1 hour to draw out the juice and dissolve the sugar. Wash, hull, and drain the whole strawberries. At serving time, stir the whole berries into the chilled, sugared berries and use the mixture between and on top of the shortcakes.

Yellow Strawberry Shortcake

YIELD: 2 9-INCH LAYERS

In my family this has always been served as strawberry shortcake—not authentic, but we like it because the cake soaks up lots of the strawberry juice and because it's tender when you bite into it.

3 cups cake flour, sifted before
  measuring
1 tablespoon baking powder
½ teaspoon salt
2 teaspoons ground nutmeg
1 cup butter or margarine (not diet
  or whipped)
2 cups granulated sugar
4 eggs
1 cup milk
1 teaspoon vanilla

Preheat the oven to 350°F.

Sift the flour with the baking powder, salt, and nutmeg. Cream the butter and sugar together until the mixture is light and fluffy. Beat in the eggs one at a time. Stir in the sifted flour alternately with the milk and vanilla, beginning and ending with the dry ingredients.

Pour the batter into two greased and floured 9-inch layer pans. Bake in a 350°F. oven 25 to 30 minutes, or until the cake is lightly browned on top and springs back when pressed with your finger. The cake should show signs of pulling away from the sides of the pan. Cool in the pans for 10 minutes, then remove and finish cooling on racks.

Strawberry Filling

10 cups sliced strawberries
1 cup granulated sugar
1 cup whipping cream
1 teaspoon vanilla extract
2 tablespoons sugar (or to taste)
1 cup whole strawberries

Mash the sliced strawberries lightly with a fork as you stir in the sugar. Refrigerate for at least an hour to allow juice to form. Whip the cream until it holds stiff peaks. Beat in the vanilla and the sugar. Wash, hull, and drain the whole strawberries.

To serve, divide the chilled strawberries in half. Fold about half the whipped cream into one-half the strawberries. Put this combination between the two layers of cake. Pour the rest of the strawberries over the top layer. Pile the remaining whipped cream on top of that and garnish with the whole strawberries.

Note: If you need a smaller amount, freeze one layer of the cake to use some other time, split the remaining layer in half and proceed as above, using half as much filling.

Spectacular Strawberry Cake

I won't even pretend that this is the sort of dessert you whip up at the last minute for Sunday night supper. It's a party dessert—great for buffets, birthday parties, anniversaries—any time you want something special and showy. You should make it a day ahead, which is a help. The fluffy strawberry filling goes into an angel food cake. Although I'm philosophically opposed to it, I've got to admit that a commercially baked angel cake would work reasonably well. Or use the recipe below.

Angel Food Cake

1 cup cake flour, sifted before
    measuring
½ cup granulated sugar
1¾ cups egg whites (about 12)
2½ tablespoons cold water
1½ teaspoons cream of tartar
1 teaspoon vanilla extract
1 teaspoon almond extract
½ teaspoon salt
1 cup granulated sugar

Preheat the oven to 350°F.

Sift the flour and ½ cup sugar together 3 times. Have the egg whites at room temperature. (Eggs about 3 days old are better here than fresh eggs.) Beat the egg whites with the cold water until they are foamy, then add the cream of tartar, flavorings, and salt, and continue beating until the whites are very stiff and glossy, but not dry. Sift the already sifted 1 cup sugar over the egg whites and fold it in, a little at a time. Next, sift in the flour and ½ cup sugar which have already been sifted together and gently fold into the egg whites, a little at a time.

Pour the batter into an ungreased 10-inch tube pan and bake 45 minutes in a 350°F. oven. Remove from the oven and cool upside down in the pan either on the legs of the pan or with the tube slipped over a soda bottle or an inverted funnel. When cool, use a knife to cut the cake away from the sides of the pan.

1½ cups mashed strawberries
1 cup sugar
1 envelope unflavored gelatin
¼ cup cold water
¼ cup hot water
1 tablespoon lemon juice
⅛ teaspoon salt
¾ cup heavy cream
2 egg whites

**Whipped cream and whole
strawberries for garnish**

Spectacular Filling

Combine the strawberries and sugar and let them stand while you prepare the rest of the ingredients. Soften the gelatin in the cold water. Pour in the hot water and stir until the gelatin dissolves. Add to the berries along with the lemon juice and salt. Chill the mixture for about 30 minutes or until it begins to set. Whip the cream until it forms soft peaks. Beat the egg whites until they are glossy and form stiff peaks. Fold the whipped cream and beaten egg whites into the strawberries.

To assemble the cake, use your fingers to pull pieces of cake from the center until you have increased the center hole to about 4 inches. Split the hollowed-out cake into two layers. Spread the strawberry mixture between the layers and put them back together again. Fill the hollowed-out center with alternating layers of filling and the pieces of cake you pulled from the center, ending with a layer of filling.

Chill overnight before serving. At serving time garnish with additional whipped cream and strawberries.

Strawberry-Peach Shortcake

YIELD: 6–8 SERVINGS

4 cups fresh whole strawberries
3 peaches, peeled and sliced
1 tablespoon lemon juice
¼ cup orange juice
2 tablespoons sugar

---

1 warm biscuit shortcake

---

1 cup heavy cream, whipped

Talk about gilding the lily! The only thing in the world I like as much as strawberries is peaches. Here they're combined in an excess of good tastes and textures. Use the butter biscuit shortcake on page 68 as the cake base.

In a large bowl combine the strawberries and peaches. Sprinkle on the lemon juice. Mash a few of the strawberries. Stir in the orange juice and sugar. Let the mixture stand about 1 hour to draw juice and form a syrup.

To serve, split the shortcake into two layers. Cover the bottom layer with about half the fruit and a small amount of the whipped cream. Put on the top layer, cover with the rest of the fruit and pour all the syrupy juice across the top. Pile on the rest of the whipped cream. Serve immediately.

Ice Cream,
Frozen Yogurt,
Ices, and Sherbet

Curly locks, curly locks,
Will thou be mine?
Thou shalt not wash dishes
Nor yet feed the swine.
But sit on a cushion
And sew a fine seam,
And feed upon strawberries,
Sugar and cream.

Nursery Rhyme

Strawberry Ice

YIELD: ABOUT 8 CUPS

The Atlanta Exposition Cookbook, *copyrighted in 1895 by Mrs. Henry Lumpkin Wilson, offered these instructions for making Strawberry Water Ice:* Mash 2 quarts berry pulp through sieve until seeds are left quite dry. Make thick syrup by boiling 1 pound sugar with ½ pint water until quite clear, adding when nearly cold to juice of berries. Lastly, add juice of 2 oranges or lemons. If this quantity of sugar does not make sufficiently sweet for general taste, more may be added before freezing.

A more contemporary recipe is surprisingly similar.

12 cups strawberries
3 cups granulated sugar
2½ cups water
Juice of 1 lemon

Wash and hull the strawberries. Mix them with the sugar and allow to stand for several hours to form juice. Puree the strawberries and sugar in a blender or food processor and strain through a double thickness of cheesecloth. Combine the juice with the water and lemon juice. Pour into the processing canister of an ice cream churn. Pack with salt and ice, then churn and harden according to the directions for your unit.

Strawberry Sherbet

YIELD: ABOUT 6 CUPS

Almost an ice, this sherbet contains no milk.

2 cups water
¾ cup sugar
1 envelope unflavored gelatin
¼ cup lemon juice
1½ cups unsweetened strawberry
   puree
2 egg whites

Mix the sugar and water in a saucepan, bring to a boil, and cook at a full boil for about 5 minutes, or until the mixture becomes syrupy. Soften the gelatin in the lemon juice and stir it into the hot syrup. Stir in the strawberry puree. Chill.

To make the sherbet, beat the egg whites until they are glossy and hold firm peaks. Fold into the strawberry mixture. Pour the sherbet into the processing canister of your ice cream churn, pack with salt and ice, then churn and harden according to the directions for your unit.

Blender Ice Cream

SUITABLE FOR UNSWEETENED
FROZEN STRAWBERRIES
YIELD: 5–6 SMALL SERVINGS

I was honestly surprised at how well this turned out the first time I made it. Nobody's going to claim it's a match for perfectly churned ice cream made in an old-fashioned freezer, but it's a lot better than most commercial ice cream and also better than poorly made home-churned ice cream. The original recipe did not call for the milk powder, so you could omit that, but I think it adds a creamy body (as well as calcium and protein) to the final product. Also, if you have a choice of appliances, I think a food processor works a little better than a blender here.

3 cups whole, frozen strawberries
   (or 1 1-pound package of
   commercially frozen
   strawberries)
½ cup whipping cream
2 eggs
⅓ cup sugar
⅓ cup non-instant, non-fat milk
   powder
1 teaspoon vanilla extract
   (optional)

Thaw the frozen berries about 10 minutes. In a blender or food processor combine the cream, eggs, sugar, and milk powder. Process for about 2 seconds, then, with the motor still running, begin to drop in the frozen berries, one at a time. Continue blending until the mixture is smooth. It will be frozen by the strawberries. Add the vanilla extract, if you are using it, in the last few seconds of processing. I suggest tasting the ice cream before adding the vanilla. You may like the flavor as is so well you decide against adding additional flavoring.

Store in the freezer 30 minutes to an hour before serving, or serve at once in chilled dishes.

Strawberry Ice Cream

SUITABLE FOR UNSWEETENED
  FROZEN STRAWBERRIES
YIELD: ABOUT 6 CUPS

2 cups sliced strawberries
½ cup sugar

2 cups heavy cream
2 cups light cream
½ cup sugar
1 teaspoon vanilla extract
⅛ teaspoon salt

Combine the sliced strawberries and sugar and refrigerate for 24 hours. The juice from the berries will combine with the sugar to form a syrup.

To make the ice cream, puree the strawberries and sugar. Combine the puree with the heavy and light cream, sugar, vanilla, and salt in the processing canister of your ice cream churn. Stir until the sugar dissolves, then pack the churn with salt and ice, and churn and harden the ice cream according to directions for your unit.

Quick Strawberry Ice Cream

YIELD: 1 QUART

2 cups half-and-half
2 cups strawberry preserves (made
  without pectin)

I got this idea for using preserves as an ice cream base from a book by Madelaine Bullwinkel, owner of a cooking school in Chicago. Gourmet Preserves Chez Madelaine is devoted exclusively to preserves and creative ways to use them.

Stir the half-and-half and preserves together. Chill for 2 hours. Pour the mixture into the processing canister of your ice cream churn. Pack with salt and ice, then churn and harden according to the directions for your unit.

Old-Fashioned Strawberry Ice Cream

YIELD: ABOUT 6 CUPS

6½ cups strawberries
1½ tablespoons lemon juice
1 cup sugar
½ cup water
3 egg yolks
2 cups light cream
Honey to taste

The proportion of strawberries to other ingredients is higher in this recipe than in the others, making the texture more dense and the strawberry flavor more intense.

Puree the strawberries with the lemon juice and refrigerate. Combine the sugar and water in a saucepan and bring to a boil. Cook and stir until the sugar dissolves and the syrup reaches 230°F. on a candy thermometer. Beat the egg yolks until light and foamy. With the mixer on high, gradually pour the hot sugar syrup into the egg yolks, beating constantly until the mixture is cooled and thickened. Stir in the cream and strawberry puree. If you want the ice cream to be sweeter, add honey to taste. Chill the mixture until you are ready to process. Pour the ice cream base into the processing canister of your ice cream churn. Pack with ice and salt, then churn and harden according to directions for your unit.

French Strawberry Ice Cream

SUITABLE FOR UNSWEETENED
FROZEN STRAWBERRIES
YIELD: ABOUT 6 CUPS

2 cups sliced strawberries
½ cup sugar

2 eggs
½ cup sugar
2 tablespoons flour
⅛ teaspoon salt
4 cups light cream (divided into 2
   equal parts)
1 teaspoon vanilla extract

This differs from the previous ice cream in that the ice cream base is cooked and chilled before freezing. Also, the strawberries are added after the ice cream has been frozen but before hardening, rather than being frozen into the mixture from the beginning. This keeps the berries from being frozen quite so hard.

Mix the sliced strawberries and ½ cup sugar and refrigerate 24 hours. The juice from the strawberries will combine with the sugar to form a syrup.

To make the ice cream, beat the eggs in a bowl until they are light and foamy. In a saucepan combine the sugar, flour, and salt. Gradually stir in 2 cups of the cream. Cook and stir over low heat until the sugar is dissolved and the mixture begins to thicken, 10 to 15 minutes.

Pour a small amount of the hot cream into the beaten eggs, stirring as you do so. Pour the egg mixture back into the hot cream in the pan, stirring constantly. Cook and stir for 1 minute. Remove from the heat and chill. Stir in the remaining cream and the vanilla. Drain the syrup from the sliced strawberries and stir it into the mixture. Pour the ice cream base into the processing can of your ice cream churn (or refrigerate until you are ready to process). Pack the churn with salt and ice and freeze according to the directions for your unit.

After the ice cream is frozen, stir in the reserved strawberries and harden in the churn or in your freezer.

Frozen Strawberry Yogurt

Frozen yogurt is wonderful for dessert or scooped on top of fruit salad. Frozen strawberry yogurt topped with some fresh strawberries makes a double helping of strawberry flavor. Of course you can make frozen yogurt from commercially produced yogurt, but it is infinitely better made from homemade yogurt and is especially good if you have home-grown berries so sweet they need no additional sugar.

2 cups strawberry puree
½–¾ cup sugar (or to taste)
1 envelope unflavored gelatin
¼ cup water
4 cups yogurt
1 teaspoon vanilla extract
  (optional)

Combine the puree and sugar and allow to stand for about 30 minutes, stirring once or twice to dissolve the sugar. Soften the gelatin in the water, then heat it until the gelatin dissolves. Stir it into the puree. Stir in the yogurt and the vanilla (if you are using it) and pour the mixture into the processing canister of your ice cream churn. Pack with salt and ice, then churn and harden according to the directions for your unit.

Frozen yogurt is also good eaten immediately after freezing, without the hardening stage, but you have to serve it at once because it is the consistency of frozen custard and will begin melting very quickly.

Frozen Strawberry-Banana Yogurt

YIELD: ABOUT 8 CUPS

1 cup strawberry puree
½ cup sugar
2 tablespoons lemon juice
1 cup mashed banana
1 envelope unflavored gelatin
¼ cup orange juice
4 cups yogurt
1 teaspoon vanilla

Adding banana to frozen yogurt contributes a uniquely smooth texture.

Combine the strawberry puree and sugar and allow to stand for about 30 minutes, stirring once or twice to dissolve the sugar. Stir the lemon juice into the mashed banana. Soften the gelatin in the orange juice, then heat it until the gelatin dissolves. Cool, then pour the gelatin into the strawberry puree. Stir in the banana, yogurt, and vanilla. Pour the mixture into the processing canister of your ice cream churn. Pack with salt and ice, then process and harden according to the directions for your unit. The frozen yogurt may also be served without the hardening stage if you do it at once. The consistency will be like that of frozen custard and it will melt quickly.

Strawberry-Yogurt Popsicles

SUITABLE FOR USING SLICED AND
SWEETENED FROZEN
STRAWBERRIES
YIELD: 12 POPSICLES

2 10-ounce cartons frozen
   strawberries, thawed
1 envelope unflavored gelatin
2 cups plain yogurt

Years ago Adelle Davis introduced many of us to the idea of popsicles made with yogurt—long before yogurt had become fashionable in America. This recipe is another version of the old Adelle idea.

Drain the strawberries and put the liquid in a saucepan. Soften the gelatin in it and then cook over low heat, stirring constantly, until the gelatin dissolves. Cool the mixture. Combine the gelatin mixture, the strawberries, and the yogurt in the processing jar of a blender or food processor and blend until smooth.

Pour into 12 3-ounce paper cups. Freeze the popsicles until they are firm enough to hold sticks upright, then insert a wooden popsicle stick in each cup and finish freezing.

Jams, Jellies, and Preserves

So you are thinking of good resolutions? That reminds us of a certain old lady reputed to be the greatest jam maker in the county. She has shelf on shelf of every kind of jam — peach, beach plum, raspberry, wild strawberry, cherry, etc. — they are there, the whole mouth-watering list. But no one ever gets to taste "Auntie's" masterpieces. She just keeps them on the shelf to be looked at. Well, those jams of hers seem to us like most of the resolutions you and I make. They're mighty pretty things, but we just keep them on the shelf.

THE FARMER'S ALMANAC
1947

There must be hundreds of recipes for strawberry jams and preserves, each just a little different from the others, but they all fall in one of four basic categories: jams, which contain mashed berries; preserves, which contain whole berries in heavy syrup; jellies, which are made of clear juice; and conserves, which include additional fruit.

Contemporary cooking technology has added commercial pectin and no-cook jams to the possibilities. Cooked pectin recipes give the highest yield, but many people feel the product tastes more of sugar than fruit; recipes made without commercial pectin produce more intense fruit flavor, but yield less for the fruit used and are less reliable for jelling. Each method has advantages in some situations.

Following a few simple rules almost guarantees success either way. First, commercial pectin is available in liquid and powdered form. They are not interchangeable. Use the kind specified in the recipe. When using commercial pectin, the berries should be *fully* ripe.

All fruit for preserving should be highest quality.

To test for the "jelly stage" in recipes that do not use commercial pectin, either use a candy thermometer and cook until it registers between 220°F and 225°F or use the metal spoon test. Dip a metal spoon into the cooking preserves, lift it out, and watch the liquid fall from the spoon. When it gets syrupy, the liquid will fall away in two drops, side by side. When the two drops begin to come together to fall from the spoon in a "sheet" you have reached the jelly stage. If you cook your jellies too much past this point you will end up with a sweet mess of tar-like consistency. Cook too little and the resulting product will be a syrup.

To store heat-processed jams and jellies (as opposed to freezer jams), use sterilized jars and screw-on tops. Paraffin is not an ideal sealer because mold can get underneath the wax and grow on top of your jelly.

Strawberry Jam

8 cups fresh strawberries
1 1¾-ounce box Sure-Jell brand
    powdered pectin
7 cups granulated sugar

This commercial pectin–based recipe gives the largest yield and most certain results.

Wash, hull, and crush the strawberries. You should have 5 cups of crushed strawberries. Place the fruit in a pan of at least 8-quart capacity and stir in the pectin. Bring to a full boil, stirring constantly. As soon as the mixture is boiling hard, stir in the sugar, a little at a time, and let the mixture return to a rolling boil that you cannot stir down. Boil hard 1 minute, stirring constantly.

Remove from the heat and immediately ladle into hot, sterilized jars. Screw on scalded lids and process in a boiling-water bath 5 minutes. If you do not expect to store the jam long, you can safely skip the 5-minute processing.

Strawberry Jelly

YIELD: 8 CUPS JELLY

9 cups strawberries
7½ cups granulated sugar
1 bottle Certo brand liquid pectin

Crushing strawberries to make jelly always seems like a terrible thing to do to good strawberries, but it's a good use for those which don't look perfect enough for other uses, and for the smaller berries you get toward the end of the season.

Wash, hull, and crush the strawberries. Bring them to a boil over medium heat and cook about 3 minutes, or just until the juices begin to flow freely. Put the mixture into a jelly bag made from 3 or 4 layers of cheesecloth and allow to hang several hours or overnight to extract the juice. You should have 4 cups of juice. Cook it down or add a bit of water to make the exact amount.

Pour the juice into a large kettle. Add the sugar. Bring the mixture to a hard boil, stirring constantly. Reduce heat and stir in the pectin. Return to a full boil and boil hard, stirring constantly, for 1 minute.

Remove from heat, skim off foam, and ladle into hot, sterilized jelly jars or canning jars. Immediately screw on hot lids and process 5 minutes in a boiling water bath. The 5-minute processing is not necessary if you do not intend to keep the jelly more than a few months.

No-Cook Strawberry Freezer Jam

YIELD: 2¾ PINTS

Although it is quick and easy, this jam is not my favorite. I think it has the taste and consistency of strawberry Jell-o, but I am including the recipe anyway because I've found that few people agree with me. Most say they like the jam because the berries taste so fresh.

4 cups ripe strawberries
4 cups granulated sugar
1 box Sure-Jell powdered pectin
¾ cup water

Wash and hull the strawberries and crush them completely, a few at a time. You should end up with 2 cups of crushed berries. In a large bowl, mix together the berries and sugar and let them stand 10 minutes. Combine the pectin and water in a saucepan. Bring to a boil and boil 1 minute, stirring constantly. Stir the hot pectin into the fruit in the bowl and continue stirring for 3 minutes. Do not worry if the sugar has not completely dissolved. Ladle the jam into freezer containers. Put the lids on immediately. Let the jam stand at room temperature 24 hours, or until set. Refrigerate to keep a few weeks, or freeze to keep up to a year.

Strawberry-Cherry Preserves

YIELD: ABOUT 3 PINTS

3 cups strawberries
3 cups sour cherries
4½ cups granulated sugar
Juice of 1 lemon
½ orange, sliced thin

Wash and hull the strawberries and pit the cherries. Combine them with the sugar, lemon juice, and sliced orange in a large, heavy saucepan. Gradually bring to a boil, then cook rapidly, stirring all the time, until the mixture is thick, about 20 to 30 minutes. Ladle into sterilized jars, screw on hot lids, and process 5 minutes in a boiling-water bath.

Strawberry Preserves

YIELD: 2 PINTS

I think of these preserves as the kind everybody's grandmother made, once-upon-a-time, simmering them in an old-fashioned kitchen that smelled nicely of the wood burning in the cookstove. The flavor is true, undiluted strawberry, and the syrup sometimes comes out a little too thin and runs through the holes in the bread—altogether an entirely authentic strawberry preserve. The pectin from the apple in the recipe doesn't change the flavor but does help set the syrup.

4 heaping cups strawberries
4 cups granulated sugar
1 *unpeeled* apple, cut in quarters
4 tablespoons lemon juice

Wash, hull, and drain the strawberries. Leave them whole. In a pan of at least 6-quart capacity, gently mix the berries and the sugar. Allow to stand for about 1 hour to draw some juice and soften the sugar.

Over low heat stir the berries and sugar, gradually raising the heat to medium high as the mixture becomes more liquid. Keep stirring. Do not let the mixture stick to the bottom of the pan. As soon as the preserves reach a full boil, add the apple quarters and boil for 10 minutes. Add the lemon juice and boil 5 minutes more. Remove the apple pieces. The syrup should be about as thick as ice-cream syrup. It will thicken more as it cools.

Ladle the berries into hot, sterilized jars, screw on hot lids, and process 5 minutes in a boiling-water bath. If you plan to use the preserves within a few months, the 5-minute processing is not necessary.

Sunshine Preserves

YIELD: ABOUT 3½ PINTS

I wouldn't make this my only way of preserving strawberries, and in damp climates where the sun doesn't shine a lot, I wouldn't try it at all, but if you live in a sunny place and have strawberries to spare, it's fun to try this old-fashioned preserving method. When it works the results are spectacular. The rest of the time you end up with slightly runny preserves.

The fruit and sugar are measured in equal amounts by weight. If you don't have scales, start with the sugar, which you can buy in pound boxes, and put strawberries in a plastic bag until it feels about the same weight as the sugar.

2 pounds strawberries
2 pounds granulated sugar
½ cup water

Wash and hull the strawberries. Put the berries and sugar in a heavy pot and let sit for about 20 minutes. Add the water and gradually bring to a boil, stirring until the sugar is dissolved. Boil hard for 2 minutes. Pour the preserves into shallow platters and set them in full sun, outside, until the liquid around the berries begins to jell. This will take 2 or 3 days. You will have to bring the platters in at the end of each afternoon before the sun goes down and put them back out each morning, and you should keep them loosely tented with cheesecloth to keep out insects.

When the mixture has begun to jell, pour into sterile jars and screw on hot, sterilized lids. You probably won't have enough preserves made this way to warrant processing in the 5-minute water bath for long-term keeping.

Strawberry-Rhubarb Jam

YIELD: ABOUT 4 PINTS

Two of the earliest fruits of spring, strawberries and rhubarb, are a classic combination.

6 cups strawberries
8 cups rhubarb cut in ½-inch
   pieces
6 cups sugar
1 lemon rind, cut in strips

Wash, hull, and mash the strawberries. Put them in a large, heavy (not aluminum) pot. Add the rhubarb. Pour in the sugar and mix thoroughly. Allow to stand at least 1 hour. Over low heat cook and stir until the sugar is dissolved. Add the lemon rind. Bring to a full boil and continue cooking until the mixture has thickened (210°F. on a candy thermometer)—about 20 to 30 minutes. Stir often to keep mixture from sticking to the bottom of the pan and burning.

Remove the jam from the heat and skim off any foam which may have accumulated. Pour the jam into hot, sterilized jars and screw on hot lids. Process 5 minutes in a boiling-water bath.

Strawberry-Honey Jam

YIELD: ABOUT 5 PINTS

I recommend treating this as a freezer jam because I think the flavors stay more true than if you can it. The flavor of the honey will be pronounced, so use one with a flavor you like.

4 cups mashed strawberries (about
   6 cups whole)
6 cups honey
Juice of 1 lemon
Rind of 1 lemon (chopped in small
   pieces)
1 teaspoon cinnamon

Combine the strawberries, honey, lemon juice, and rind in a large, heavy pan. Bring to a boil and boil hard for about 20 minutes, stirring almost constantly to avoid sticking and burning. When the mixture is thickened, remove from heat, stir in the cinnamon, and pour into jars or freezer containers. Seal with hot lids or cool completely and freeze.

Beverages

A pot of
Strawberries gathered in the wood
To mingle with your cream.

Ben Johnson
1603

Strawberry Breakfast Drink

SUITABLE FOR UNSWEETENED
  FROZEN STRAWBERRIES
YIELD: 1 TALL DRINK

1 egg
¾ cup milk
1 tablespoon nonfat milk powder
1 tablespoon nutritional yeast
1 teaspoon vanilla extract
½ cup whole fresh or frozen
  strawberries
Sugar to taste
Nutmeg to taste
Strawberries for garnish

Starting the day off with this drink will fortify you the way Popeye's can of spinach did him in the early cartoons.

Combine all ingredients but nutmeg in the blender and blend until thick and frothy. Pour into a tall, chilled glass and garnish with nutmeg and a strawberry.

Strawberry Cooler

SUITABLE FOR SLICED AND
  SWEETENED FROZEN
  STRAWBERRIES
YIELD: 6 SERVINGS

1 10-ounce package frozen
  strawberries or 1 cup fresh sliced
  strawberries and ¼ cup sugar
1 cup milk
1 cup water
½ teaspoon vanilla
6 ice cubes

Here's a quick, cool, pick-me-up for a hot summer afternoon or, if you use frozen berries, a reminder of summer any time of year.

Combine all ingredients except the ice cubes in the processing jar of a blender or food processor. Blend until the mixture is smooth, then add the ice cubes, one at a time, and continue processing until they have been thoroughly incorporated. Serve immediately.

Grandmother's Teapot Juice

YIELD: 4–6 SERVINGS

1 cup fresh whole strawberries
Juice of 2 lemons
2 cups sugar
4 cups cold water (spring water if you can manage it)
Whole strawberries or fresh mint leaves for garnish

Only if you have very sweet, full-flavored strawberries would this be worth the trouble. But if you do have them, the resulting beverage is heavenly.

Crush the strawberries and mix them with the sugar and lemon juice. Let the mixture stand at least 3 hours. Stir in the water and strain through a fine sieve lined with a single thickness of cheese-cloth. Serve over crushed ice or ice cubes in tall glasses. Garnish with additional fresh berries or fresh mint leaves.

Other Uses. Use this cocktail as the liquid to make strawberry gelatin with unflavored gelatin. Use as part of the liquid in lemonade and in mixed drinks. Or bring to a boil and simmer 10 minutes with a cinnamon stick, lemon slices, and a few whole cloves to make a hot, mulled strawberry drink.

Strawberry Apple Cooler

SUITABLE FOR UNSWEETENED
   FROZEN STRAWBERRIES
YIELD: 6 SERVINGS

4 cups apple juice
1⅔ cups strawberry puree
Crushed ice

Who needs soda pop if they can have this?

Mix 2 cups of apple juice with the strawberry puree and mix in the blender until frothy. Add the rest of the apple juice, stir, and pour into glasses about half-filled with crushed ice.

Nan Marlowe's Strawberry Lemonade

YIELD: 1½ QUARTS (6–8 SERVINGS)

3 cups water
1½ cups sugar
6½ cups fresh strawberries
½ cup lemon juice

You can make this in seconds any time you have strawberries if you've already cooked and cooled the sugar syrup.

Cook the water and sugar together until the sugar dissolves. Cool and chill. To make the lemonade, puree the strawberries in a blender or food processor, combine with the lemon juice and sugar syrup, and serve with ice cubes.

Strawberry Ice Cream Soda

YIELD: 1 LARGE SERVING

⅓ cup strawberry puree (sweetened to taste)
¼ cup milk
1 tablespoon nonfat dry milk powder (preferably not instant)
1 scoop homemade strawberry or vanilla ice cream
Seltzer water or club soda (cold)

Dare I say the word "nutrition" out loud? Here's a chance to slip vitamin C, calcium, and protein into a person who's aware only of getting a soda-fountain treat.

Put the puree in the bottom of a tall glass. Dissolve the milk powder in the ¼ cup milk and mix into the puree. Add the ice cream. Slowly pour in the carbonated water and stir gently just to mix. Serve with two straws.

Strawberry Punch

SUITABLE FOR USING WHOLE
  FROZEN STRAWBERRIES
YIELD: 4 QUARTS (16–24 SERVINGS)

3 cups strawberries
6 cups water
1½ cups sugar
1½ cups honey
Rind of 1 orange
Rind of 1 lemon
4 cups water
1 cup orange juice
1 cup lemon juice
½ cup lime juice
Lemon slices for garnish
1 cup fresh or frozen strawberries
  for garnish

I adapted this recipe from one I found in an old Ladies' Home Journal Cookbook *that I have had for more than 20 years. In fact it was the first cookbook I ever owned. I still turn to it for ideas. The punch is quite a bit of work, but absolutely lovely for a special-occasion buffet or to carry in a big cooler jug for a picnic.*

Combine the strawberries and 6 cups of water in a saucepan. Bring to a boil and simmer for 5 minutes. Strain the juice through a sieve and chill. Combine the sugar, honey, and orange and lemon rinds with the 4 cups of water in a saucepan, bring to a boil and simmer 5 minutes. Cool and refrigerate.

To serve the punch, mix together the strawberry juice, honey syrup, and citrus juices. Stir in the lemon slices and strawberries. Pour over ice in a punch bowl or pitcher.

Strawberry Milkshake

YIELD: 2 SERVINGS

Try a milkshake for dessert instead of pie or cake at lunch or after a light supper. Even people who say they don't like milk usually enjoy milkshakes. If you need to serve several people, it is better to mix up milkshakes no more than two at a time than to try to do large quantities. They blend more evenly in small amounts.

2 tablespoons strawberry jam or preserves
1 cup homemade vanilla or strawberry ice cream
1 cup cold milk
1 tablespoon nonfat dry milk powder (preferably not instant)

Blend all ingredients in a blender or food processor until almost smooth. Leave a few little lumps of ice cream in the mixture. Serve with a straw.

Strawberry Daiquiri

SUITABLE FOR UNSWEETENED FROZEN STRAWBERRIES
YIELD: 1 DAIQUIRI

This is definitely a party drink; I always envision strawberry daiquiris being served to elegantly dressed people gathered in the early evening on a lush green lawn.

1 jigger (1½ ounces) light rum
4–6 whole fresh or frozen strawberries
1 tablespoon fresh lime juice
1 teaspoon sugar
4–5 ice cubes

Combine the rum, strawberries, lime juice, and sugar in the blender. Process until the strawberries are pureed. With the motor still running, drop in the ice cubes one at a time and blend until the mixture is slushy.

Strawberry-Banana Drink

SUITABLE FOR SLICED AND
  SWEETENED FROZEN
  STRAWBERRIES
YIELD: 3 LARGE OR 6 SMALL
  DRINKS

2 cups fresh or 2 8-ounce
  packages frozen strawberries
2 bananas
1 cup yogurt
1 tablespoon nonfat dry milk
  (preferably not instant)
½ cup milk
2 tablespoons honey (omit if using
  sweetened strawberries)
¼ teaspoon vanilla extract

If using frozen strawberries, thaw slightly. Combine all ingredients in the blender or food processor and blend until smooth.

Strawberry Comfort

SUITABLE FOR UNSWEETENED
  FROZEN STRAWBERRIES
YIELD: 2 4-OUNCE OR 1 8-OUNCE
  DRINK

1 jigger (1½ ounces) Southern
  Comfort
4–6 whole frozen strawberries (not
  thawed)
¼ cup orange juice
4–5 ice cubes

I think this is what would be called a "lady's drink" by professional bartenders. It's pretty, sweet, and light.

Combine the Southern Comfort, strawberries, and orange juice in the blender. Blend until the strawberries are pureed, then drop in the ice cubes one at a time with the motor still running. Blend until slushy.

Serve garnished with a fresh strawberry.

Growing
Strawberries

The barberry, respies,
and gooseberry too,
Look now to be planted as other things do.
The gooseberries, respies and roses all three,
With strawberries under them trimly agree.

Hide strawberries, wife,
To save their life.
If frost do continue, take this for a law
The strawberries look to be covered with straw.

Thomas Tusser
FIVE HUNDRED POINTS
OF GOOD HUSBANDRY
1557

If you set your mind to it, you could probably buy fresh strawberries—from Florida, California, Texas, Mexico—about nine months of the year. It's easy to take them for granted in the same way we expect to find iceberg lettuce, lemons, and carrots at the produce counter of the grocery store any time of year. And unless you're near a good garden it's easy to start believing that the way the supermarket version of all those fruits and vegetables, including strawberries, tastes is the way they are *supposed to taste*. Of course it ain't so. Not for lettuce. Not for carrots. Not for strawberries.

Home-grown strawberries differ more dramatically from those available commercially than any other produce I can think of—except possibly fresh peas. A well-managed home strawberry patch produces berries that are so sweet they don't need sugar, so flavorful they don't need accompanying ingredients to perk them up, and so juicy you want to eat each berry all in one bite to keep any of the good stuff from dribbling down your chin. Strawberries from a pick-your-own place come in a close second because you can get them at their peak. What keeps them from being quite as good as those you grow yourself, in spite of their freshness, is that such places generally choose varieties for their firmness, large size, and heavy production, with flavor a secondary consideration. For flavor, the best berries are almost never the firmest ones and only sometimes the largest ones. Those with good juice and flavor tend to be a little too tender for standing up to the hordes of pickers who work through pick-your-own fields. Also, such fields are usually huge—measured in acres rather than a few square feet. Most growers can't or don't work as much compost and other organic material into the soil of such large plots as the home gardener does into a small, personal patch. You can detect the difference in flavor.

I used to think growing your own strawberries was strictly for those with enough land to lay out a good-sized traditional patch. Then I moved to California and lived for a time in one of those developments where the houses were so close together that, were it not for the fences between them, we could have reached out and touched hands from our bedroom windows. Not only that, the area I was in was essentially desert. Whatever grew there grew in soil carried in to ameliorate the adobe, and was irrigated by water piped in from distant rivers. In that unlikely situation my kids brought me basket after basket of splendid strawberries from a friend whose mother said she had too many and was tired of making preserves.

This I had to see! I paid her a visit and found that in her handkerchief-sized back yard, she had created raised beds for some prize shrubs and had planted strawberries around the edges as border plants. Runners hung over the landscaping ties dripping with berries. Yet once she had created the beds and planted the strawberries, about the only care they got was the same routine watering and feeding with which she sustained her shrubs. Without thinking much about it, she was doing everything right. Her fertilizing program included using compost and other organic material as mulch to slow the speed with which precious water evaporated into the desert air. The warm sunny days and cool nights to which her shrubs responded so well are just what strawberries need to develop the best flavor and highest sugar content.

Since then I've paid more attention to the possibilities for growing strawberries and discovered that there's a variety and a method suitable for almost any situation. But being able to grow strawberries is hardly a modern-day accomplishment, either.

The *Booke of Cookery*, known to have been in the hands of Martha Washington as early as 1749, instructs readers on how to make strawberries bear early: *Water yr straberries once in three dayes with water wherein hath been steeped sheeps dunge or pigeons dunge, & they bear much earlier. Plants waterd with warme water will come up sooner & better than with cold water....*

That conforms closely with what we know today about growing strawberries. They like well-drained soil that is moist but not wet, warm ground, and lots of organic matter. Martha's book doesn't mention it, but they also like mellow, well-prepared soil—not newly plowed sod—with a pH between 5.0 and 6.5.

The best site is one with a gradual slope that drains well but isn't steep enough to make erosion a problem. If you are in an area where late frosts endanger early crops, avoid a low spot where the cold air will settle down and nip your strawberry plants. If your aim is the earliest berries, the site should slope toward the south. If you're one of those rare, patient people willing to wait a bit for later berries, find a north-sloping site. If you can manage one of each you'll extend the strawberry season for your garden significantly.

But let's not make too much of sites and slopes. If you live on a mini-lot with more deck space than ground, you can still grow strawberries in built up beds, planters, tubs, even window boxes, as long as you can provide a spot that is sunny most of the day. I'll tell you more about container growing and raised beds a little later. Let's cover the basic garden growing information first, because what you do with containers is basically just an adaptation of garden techniques.

IN THE GARDEN

In a garden plot, experts recommend beginning to work and enrich the soil as much as three years ahead of time with a green manure crop and compost. The main thing is not to plant in newly tilled soil because of the white grubs and wireworms. For more extensive information on what to plant as a green manure crop and on composting to build soil that will give you the best strawberry crop possible, the Garden Way Publishing bulletin *Grow the Best Strawberries* is a useful resource packed with information. Also, you might want to order Farmers' Bulletin No. 1043, *Strawberry Varieties in the United States*, from the U.S. Department of Agriculture. Write: Superintendent of Documents, U.S. Government Printing Office, Washington DC, 20402.

These bulletins will give you lots of information on strawberry varieties, which seem to be available in mind-boggling numbers. At least twenty-seven varieties are in regular use by commercial growers. Burpee's catalog offers twenty different varieties for home gardeners. This is not actually as overwhelming as it seems at first because certain varieties are for specific climates. For instance, in the northern states where plants have to withstand winter cold, Midway, Sparkle, and Catskill are widely grown, while in the South, growers choose varieties that do not need a winter rest period—Florida Ninety, Blakemore, Albritton, and Headliner, for example. Pocahontas, Dixieland, and Catskill often are planted for their resistance to frost. If you buy your plants at a local nursery, the people who work there should be able to help you with your choice, if indeed they have more than one or two varieties. If the nursery is reliable, the varieties they carry will be those that have historically done well in your area. If you order plants through a catalog, the descriptions will tell you what climate each variety is suited to, and your local agricultural extension agent can make suggestions as well.

Besides climate and growing conditions, you will have to choose between June-bearing and everbearing plants. The June bearers are those which, just as their name implies, produce most of their berries in June. They give you lots of berries within a short time. Everbearing strawberries produce a crop in the spring and another later in the summer or fall. Neither the spring nor the fall crop will be as large as the typical spring crop of the June bearers. Each has its advantages. If you are especially interested in freezing and preserving, it's nice to get the berries all at once so you can put up significant quantities and get the job done before the rest of your garden starts demanding attention. If you prefer a steady supply of fresh berries and are less concerned with putting them up, the everbearing plants will probably be a better choice. Perhaps the best known of the everbearing varieties is Ozark Beauty.

Yet another kind of strawberry is the alpine strawberry, which has a flavor similar to that of wild strawberries, and is a perennial plant which does not produce runners. It is good for growing in containers and ornamental borders.

You can now order strawberry seeds for alpines as well as for some experimental everbearers. Obviously, it takes *much* longer to get a strawberry from a seed than from a plant—at least 90 days longer, but it's a much less expensive way of getting your stock and fun if you enjoy experimenting.

Having prepared your soil and acquired your plants, you are ready to plant. In most areas, early spring is the recommended planting time. In the South, late fall is also popular. The three standard ways to plant strawberries are: in hills, in spaced matted rows, and in solid matted rows.

In the hill system you remove the runners to cultivate the mother plant as fully as possible. This produces larger berries, but you don't get as many of them. With hills, set plants 10 to 12 inches apart.

In the spaced matted-row system you train the runners so that they grow 6 to 8 inches apart by covering each runner with soil to root it as soon as it begins to grow. This is a very attractive-looking, neat way to grow strawberries and it produces good-sized berries. It's also a lot of work. Set plants about 1½ feet apart each way in rows 4 feet apart. Twenty-five plants will cover about 38 feet of row.

The solid matted-row system allows the runners to spread and root at will. All you have to do is trim off any runners that stray into the walkways between rows. It's easy and produces lots of strawberries, but they will be smaller. In the fall, plants need to be thinned to allow three or four inches between plants. Space plants about 1½ feet apart in all directions. Twenty-five plants will cover about 140 square feet.

Louise Riotte's simple directions from *Grow the Best Strawberries* make planting easy:

1. Soil test should have determined whether you should add lime and fertilizer, and how much of either. If soil hasn't been tested, a rule is to add a 12-quart pail of a commercial fertilizer such as 5-10-10 per 1,000 square feet.

2. Till or rake soil several times in two weeks prior to planting. Each time you do this you will eliminate many freshly germinated weeds. These are weeds that will never rise to cause trouble in your new strawberry bed.

3. You're ready to plant. Trim off most of the old leaves from each plant. One can be left on, if it makes handling the plants easier for you.

4. Thoroughly soak the plant roots. Now place them in a basket, bucket, or sack, so they will not dry out. When planting, never remove more plants from basket than you can put into the soil in 15 minutes.

5. Use a trowel, a dibble, or some other tool for making holes. This can be done quickly by inserting blade of trowel into earth, then pressing it back, and tipping it to both sides. Hole will be large enough for spread of roots.

6. Set plants at correct depth. The base of the crown should be at the level of the soil surface. Plants set too deep will smother and die; and if they are set too high, they will dry out.

7. Spread out roots, then carefully firm the soil around the roots. Take care with this step, for the success of the planting depends on it. Leave no air pockets in the soil.

8. If soil is dry, pour a pint of water or soluble fertilizer around each plant. This is excellent insurance to make certain the plant roots don't dry out, which would cause the plant itself to die.*

About six weeks after planting, apply a side dressing of a balanced garden fertilizer or a manure or compost tea along the lines of that described by Martha Washington's book.

Now, here's the hard part. When the blossoms appear on your plants the first year you should *pick them off so that they don't set fruit!* Isn't that just awful? All that work and you have to wait another whole year for strawberries. But there's a good reason and it's worth living through the wait. The second-year berries will be larger and more plentiful because you let the plant strengthen without fruiting that first season. Console yourself—if you do let strawberries fruit the first season you get a disappointing crop.

Of course as the gardening season progresses you'll keep the strawberries cultivated and weeded, even if you can't have berries just yet. And you'll mulch those plants. Mulch may be the most important thing you do for your strawberries. It keeps in moisture, holds down weeds, protects plants against cold, helps retain soil, keeps berries clean—and looks nice. Use whatever materials are most readily available in your area—straw, pine needles, strawy manure, hay, even black plastic mulch.

CONTAINER GROWING

Strawberries grow astonishingly well in almost any container. I've even seen lovely hanging baskets of strawberries, with runners trailing over the sides laden with strawberries and blossoms. But in such small containers, the strawberries are more decorative than a source

*From *Grow the Best Strawberries*, Garden Way Publishing Bulletin A-1.

of food because you won't produce enough for more than a few servings. The same is true of the traditional clay strawberry jar and those attractive barrels with holes cut in the sides. To grow strawberries in crop quantities requires larger containers.

The literature on container growing tends to get a little wifty on the subject of suitable containers. I am thinking, for instance, of the suggestion that you grow strawberries in an old bathtub with extra holes in the bottom for drainage, and I can't help but note that if you have room for an old bathtub, you probably have room to plant your strawberries right in the ground. Then too, I have visions of a couple of apartment dwellers trying to lug an old tub (those things are heavy!) into the elevator in such a position that the door will still close. And they haven't even gotten to the problem of putting holes in the bottom of the tub yet. But, no question, if you put your mind to it, you *could* grow strawberries in an old bathtub.

Among the more practical possibilities are window boxes, half-barrels, and the redwood plant "towers" now being sold through the major seed catalogs especially for growing on patios and other small spaces. The 2-foot high tower will hold fifty plants; the larger, 4-foot high tower creates 46 feet of row and holds nearly twice as many plants.

Another successful small-space arrangement is the pyramid or ring garden. These, too, are sold through seed catalogs, but you can easily make your own. In the standard size, the base ring is six feet in diameter. Two more rings, each smaller than the previous one, top the first ring. You fill the first ring with planting mix, place the second on top of that, and then repeat with the smallest ring. If you are placing the rings right on the ground, no base is necessary. But you would still treat the project as container growing because you would need to create and put in potting mix and because the raised rings would keep the strawberry roots from ever reaching the ground base.

On a patio or porch, you can improvise a successful vertical strawberry patch by building an oblong cage or a column of chicken wire, lining it with sphagnum, and planting strawberries all over the exterior. To do it, fill the column from the ground up, pulling the roots through the chicken wire and sphagnum lining, and filling the center with potting mix as you go.

Everbearing varieties are recommended by some experts for growing in containers, but I've noticed that the Burpee people offer a combination of Superfection (everbearing) and Royalty (June-bearing) plants for planting in their redwood tower.

For window boxes and similar containers try everbearers. For hanging baskets, consider the smaller alpine berries, grown from seed.

Because of the variety of containers and locations in which you might try growing strawberries, each project has to be thought of as experimental. The keys to success are that the container receive eight or more hours of sun each day (although alpines in hanging baskets do well in a partly shaded spot), that the potting mix be rich and humusy and drain well, and that you keep the plants constantly moist. The potting mix can contain as much as one-third garden loam, provided it is not heavy clay. A standard mix for growing food in containers is: *seven parts loam, three parts peat moss, and two parts coarse sand or perlite, enriched by adding ¾ pound of a whole (complete) fertilizer and three ounces of lime for each bushel.*

You should fertilize your strawberries with manure tea or compost tea as soon as they begin to grow.

Louise Riotte writes that if you plant everbearing varieties in a container such as a window box and pinch off the runners, you can expect up to 200 berries from each plant, given proper care and conditions. Of course this figure would change if you use a cage or tower and root the runners, but you can see that the yield from container-grown strawberries can definitely be worth the effort.

A Short History of the Strawberry

In Elizabethan English folklore, the belief was held that proximity to other plants could affect a plant to such an extent that they "imbibed each other's virtues and faults." The strawberry, however, was supposed to be the exception to this rule, and was said to be able to thrive in the midst of "evil communications without being corrupted."

Anthony S. Mercatante
THE MAGIC GARDEN

A bachelor I once knew waxed so enthusiastic about matrimony when he finally married that his friends said he talked as though he had invented it himself. I think much the same is true of our current enthusiasm for strawberries. We may not think we invented them, but surely we discovered their value, and certainly we've improved them. We know they've been around for a while, but the experiments in growing, the appreciation of their nutritional value, and the many imaginative ways of serving strawberries—these must be new to contemporary culture. Aren't they?

Well, it has been reported that the Roman senator Cato, who lived from 234 to 149 B.C., loved strawberries soaked in wine or seasoned with salt and spices so much that, rich and powerful as he was, he got up early each morning to inspect his strawberry patch and remove insects that threatened the plants.

A Saxon plant list of the tenth century includes the strawberry; in 1265 the "Straberie" is mentioned in the household writings of the Countess of Leicester. By 1580 Thomas Tusser, author of *Five Hundred Pointes of Good Husbandrie* was giving directions for domestically cultivating strawberry plants from the woods:

> *Wife, into thy garden and set me a plot*
> *With strawberry roots, of the best to be got:*
> *Such growing abroade, among thorns in the wood*
> *Well chosen and picked, prove excellent good.*

I'm not crazy about his division of labor, but his advice was sound. In those days I guess the objective was to be the writer of gardening directions, not the wife who carried them out. Interestingly, Tusser also discussed using strawberries as ground covers in shrub beds, much as my neighbor in California grew strawberries in the raised beds around her prize shrubs.

Thomas Hill must have had a wife, too, because in *the Gardeners Laybrinth* in 1577 he wrote, "Strawberries require small'labor, but by dilligence of the gardener becometh so great, that the same yieldeth faire and big Berries..." Hill also wrote that strawberries are good with cream but better with wine, a treatment that is reflected in a medieval recipe for strawberries directing the cook to "Take Strawberys, & wash them in... gode red wyne..." The popularity of strawberries in wine persists. The classic *Larousse Gastronomique* gives a recipe for strawberries marinated in champagne, and many contemporary recipes for strawberries in red wine, white wine, and garnished with port wine gelatin come remarkably close to those early recipes.

Similarly, a contemporary classic, Coeur a La Creme, a heart-shaped creamy cheese dessert served surrounded with fresh strawberries, seems to have had its origins in a medieval recipe using rennet (known as junket today) to set milk, after which it was drained in reed baskets, to produce a "cream cheese" to serve with strawberries. Even today, authentic recipes for Coeur a la Creme instruct you to mold the cheese in heart-shaped baskets especially made for the purpose.

As for our appreciation of the nutritional attributes of strawberries, their being full of vitamin C and iron but low in calories—nothing new there either. In 1324, an ancient hospital in the north of France was cultivating strawberries at considerable expense as part of a medicinal herb garden. It is true that from the sixteenth century well into the eighteenth century, some respected scholars believed strawberries were poisonous because they grew close to the ground where they could be contaminated by the urine of snakes and toads. That notion was finally laid to rest when the Swedish botanist, Carl Linnaeus, cured himself of gout on a strawberry diet in the late 1700s. Since then, at various times and places, a tea made from strawberry roots and leaves has been used to treat diarrhea, while strawberry juice was recommended for whitening skin and teeth and easing sunburn. Today's strawberry-based cosmetic creams and conditioners have a legitimate history.

But I find most fascinating of all the diversity of experimenting with cultivation that shows up from the oldest written records about strawberries. The earliest experiments, of course, were to find ways of cultivating domestically the plants found in the woods so that Milord's palate wasn't entirely dependent on the fruits of foraging. Apparently cultivation improved the taste of the strawberries, too, because in his book *The Right Way to Long Life*, a physician in Bath, Tobias Venner, wrote not only about the strawberry's medicinal qualities, but also about its delightful flavor when cultivated, saying that wild strawberries weren't as good as those "manured in gardens." That was in 1638. Also in England in the 1600s, Sir Hugh Plat grew strawberries and described in detail how to get them to ripen early or to delay their ripening, how to grow bigger berries and more of them, how to produce a double crop, and how to protect the plants against frost. The advice included planting on a south-facing bank, cutting back the plants as soon as the first crop was done, and fertilizing with "cow or pigeon dung." Sounds familiar, doesn't it?

In a book entitled *A History of the Strawberry: From Ancient Gardens to Modern Markets*, Stephen Wilhelm and James E. Sagen trace the development of strawberry varieties from the first mentions in historic writings through varieties current in 1974. The authors give an astonishing account of strawberries as they developed into: the strawberry of Canada, the strawberry of Virginia, the *scarlet* strawberry of Virginia, the strawberry of Chile, the Californian, the Hawaiian Chilean, and the Pine strawberry. The book follows the appearance, disappearance, and reappearance of strawberries all over the world, piecing together the story from fragments of old research and writing. I think the chapter entitled, "The Strawberry Returns to America" sounds wonderfully dramatic. A truly overwhelming piece of research, the book is fascinating enough that I've sat around airports reading from it when I could have been reading the latest James D. McDonald mystery.

Wilhelm and Sagen tackle the question of how strawberries got their name with earnestness and erudition, disagreeing with the popular theories that it was because strawberries are traditionally mulched with straw or because European children used to thread them onto straws and sell them in the streets by the straw. By tracking down when the word first appears in writing, they conclude that strawberries were so named because of the tendency of the mother plants to *strew* baby plants about on runners—thus, we're really eating "strewberries."

On top of that, they're not really berries at all—technically not even fruit, but the enlarged ends of the stamen. Never mind. Let's just go with Izaak Walton. In *The Compleat Angler* he wrote, "Doubtless God could have made a better berry, but doubtless God never did."

Bibliography

Grieve, Mrs. M. *A Modern Herbal*, Vol. II.
New York: Harcourt, Brace & Company, 1931.
This two-volume herbal, now available in paperback from Dover Publications, Inc., contains more information than most of us could possibly absorb about the culture and medicinal and cosmetic uses found in folklore and science for practically every plant you can think of.

Hess, Karen. *Martha Washington's Book of Cookery*.
New York: Columbia University Press, 1981.
Karen Hess is an outstandingly good historian, cook, and food historian. Her introduction and commentary for the cookbook which once belonged to Martha Washington give us a full sense of what foods were eaten and how they were prepared in America in the 1700s.

Hieatt, Constance B. and Sharon Butler. *Pleyn Delit: Medieval Cookery for Modern Cooks*. Toronto: University of Toronto Press, 1976.
Hieatt and Butler have put together a book of medieval recipes followed by contemporary modification for modern cooks, which, along with their lucid commentary, make it possible to taste what medieval food was like and to understand how today's cooking has evolved from the old recipes.

Riotte, Louise. *Grow the Best Strawberries*. Pownal, VT: Garden Way Publishing, 1977.
A concise, to-the-point little publication from the Garden Way Country Wisdom Bulletins series tells you everything you need to know to grow strawberries. It maybe ordered by writing Garden Way Publishing, Schoolhouse Road, Pownal, VT. 05261.

Wright, Michael. Ed. *The Complete Indoor Gardener*. New York: Random House, 1974.
This book contains much useful information on container growing inside and outside.

Wilhelm, Stephen and James E. Sagen. *A History of the Strawberry; From Ancient Gardens to Modern Markets*. Berkeley, CA: University of California Division of Agricultural Sciences, 1974.
The history of strawberries seems like such an unlikely subject that I almost didn't look at this book at all. I'm glad now that I did because the information is fascinating and overwhelming in its thoroughness. The full address for ordering the book: Agricultural Publications, University of California, Berkeley, CA 94720 or Strawberry Hill Press, 2594 15 Avenue, San Francisco, CA 94127.

Index